T0228693

MONOCHROM'S ARSE ELEKTRONIKA ANTHOLOGY

Editors: Johannes Grenzfurthner, Guenther Friesinger, Daniel Fabry
Publisher: RE/SEARCH
Copy editing: Melinda Richka
Layout: Daniel Fabry, Anika Kronberger
Arse Elektronika logo: Tokyo Farm
RE/SEARCH Staff: V. Vale, Marian Wallace, Seth Robson, Robert Collison, Jared Power, Alex Lavine, Cayla Lewis, Stellar Kutchins, Michael Raines, Joanna Sokolowski, Ilana Fried, Joe Donahoe

This publication was supported by the Department of Art Funding / City of Vienna, Austria.
ISBN USA (RE/SEARCH): 978-1-889307-30-5
ISBN EU (edition mono/monochrom): 978-3-902796-02-8
Printed in Poland

LETTERS, ORDERS & CATALOG REQUESTS TO:
RE/SEARCH PUBLICATIONS
20 ROMOLO #B
SAN FRANCISCO, CA 94133, USA
PH (415) 362-1465
info@researchpubs.com
www.researchpubs.com

Arse Elektronika 2009 was organized by monochrom.
www.monochrom.at/english

Curator: Johannes Grenzfurthner
Financial supervisor: Guenther Friesinger
Corrections of all kinds: Evelyn Fuerlinger
Web supervisor: Franz Ablinger
Press: Roland Gratzer
Web design: Anika Kronberger
Technical supervisor: Daniel Fabry
Co-organizers: David Fine, Carol Queen

Many thanks to the supporters of Arse Elektronika 2009: Laughing Squid, Center for Sex and Culture, Chaos Computer Club, Simone Davalos, David Calkins, Melinda Richka, Mediapathic Steen and the Department of Art Funding, City of Vienna, Austria.

CONTENTS

OF INTERCOURSE AND INTRACOURSE
SEXUALITY, BIOMODIFICATION AND THE TECHNO-SOCIAL SPHERE

Scottish SF author Iain Banks created a fictitious group-civilisation called "Culture" in his eponymous narrative. The vast majority of humanoid people in the "Culture" are born with greatly altered glands housed within their central nervous systems, who secrete - on command - mood- and sensory-appreciation-altering compounds into the person's bloodstream. Additionally many inhabitants have subtly altered reproductive organs - and control over the associated nerves - to enhance sexual pleasure. Ovulation is at will in the female, and a fetus up to a certain stage may be re-absorbed, aborted, or held at a static point in its development; again, as willed. Also, a viral change from one sex into the other, is possible. And there is a convention that each person should give birth to one child in their lives. It may sound strange, but Banks states that a society in which it is so easy to change sex will rapidly find out if it is treating one gender better than the other. Pressure for change within society would presumably build up until some form of sexual equality and hence numerical parity will be established.

Does this set-up sound too futuristic? Too utopian? Too bizarre?

We may not forget that mankind is a sexual and tool-using species. And that's why monochrom's conference Arse Elektronika deals with sex, technology and the future. As bio-hacking, sexually enhanced bodies, genetic utopias and plethora of gender have long been the focus of literature, science fiction and, increasingly, pornography, this anthology sees us explore the possibilities that fictional and authentic bodies have to offer.

Our world is already way more bizarre than our ancestors could have ever imagined. But it may not be bizarre enough. "Bizarre enough for what?" – you might ask. Bizarre enough to subvert the hetero-sexist matrix that is underlying our world and that we should hack and overcome for some quite pressing reasons within the next century.

Don't you think, replicants?

Johannes Grenzfurthner / monochrom

Eleanor Saitta

DESIGNING THE FUTURE OF SEX

Introduction

Objects, experiences, and systems are all products of design and have a huge effect on every aspect of our lives, in ways we often do not even realize. Sex and relationships are, of course, no exception. Social desires feed technological changes as they fund, prioritize and permit research, but the objects and concepts that come from those changes affect society too. Can we subvert those desires through design?

Social change requires mass disruption – individual hacking can at best prefigure and rehearse it. When technology changes society, the change starts slowly and accelerates and accumulates as it goes. Five years of change barely registers, but twenty can change society to the point where it is almost unrecognizable. Compounded over time, design can be a profoundly disruptive social force. Knowing what might be later opens our eyes to what can be now, and the fictional future can teach us about what is really transgressive in the actual present – it permits us to see our context from the outside.

It's going to get pretty weird around here.

In this paper, we will explore this territory using design fictions as a lens. Specifically, we are going to examine some of the following questions, which were used to generate the fictions we will see:

Intimacy
- How can artifacts and services enable new forms of intimacy, and in doing so, alter the constraints under which it acts?
- Does virtualizing intimacy help or hurt real, modern relationships?
- Can we qualitatively change the kind of sex we are capable of?

Identity
- Can redesigning the experience of sexual interaction profoundly change the interactees?
- Can an artifact alter our socio-sexual identity?

Awareness
- How does making difference visible change society?
- Can we change sex by intermediating it?
- Can mediated social awareness empower networks of sexual change?

Control
- How can design shift the locus of control in the interactions that revolve around sex?
- Can we create entire new vectors of sexual control, or erase old ones?

Economics
- Can we enable in people the power to escape rigid heteronormative economic relationships?
- Can designed changes to sexual economics act as a disruptive social force?

A Note on Politics and Possibility

The fictions in this paper tread right around the line of possibility. In general, everything we are considering is firmly in the realm of the known possible – often not something we can realize immediately, with some notable exceptions, but not so far out on the development curve that we cannot understand it – the near future, where design becomes meaningfully possible. To understand how far out we may be looking, each fiction will note an approximate timeline of when it might be meaningfully possible. To the extent that the distinction matters, the fictions here are primarily all about designed objects, environments, or systems, even if they are very tightly integrated with the body. We will mostly avoid the territory of genetic engineering and similar issues.

It is worth noting here that just as many of the questions we are asking are not directly sexual, or even purely relevant to intimate relationships, many of the fictions are not either – if for no other reason than that sex is both necessarily political and heavily embedded in its lived context. It is this more complicated territory that we address here.

Some of these fictions are intended as positive statements, but many of them are more ambiguous. The world will change, whether or not we really want it to.

Fictions

Modern Flagging
Relevant to: Identity and Awareness
Timeline: 5 years

Fractured identities complicate meeting people in a situated manner. Things like flagging (from the BDSM world) do not really generalize across social groups, but personal ad sites at best add another layer of division – after all, other people have to opt in to the site. The usual mechanisms of subculture work, but they lack any kind of geographic referent – the people you see at the club across town are not the people you see at the corner store. How do you find out if the woman next door is into the same things as you? Subcultures and sexual diversity can alienate us from our surroundings even when they do not need to, making the lack of commonality the default assumption.

If you believe privacy is dead, why wait for people to announce they are looking?

You walk into a grocery store. Your phone looks around, figures out whom all the other phones belong to, and then drift-nets through their data shadows. You get notified, while picking out tomatoes, that the cute woman by the fruit really likes getting choked. You go say "Hi."

The mobile parts of a system like this are mostly trivial – the only hardware support needed is a bit more locative space-awareness and phones that are a bit more promiscuous with identities. Filtering the data shadow is the interesting part – semantic web style analysis, social network enumeration, heuristic analysis of past interactions ("If you liked sleeping with Susan, you'll love Mary!"), and careful use of all of the (increasing amount) of relevant information people put online.

There is an argument, of course, that there is a standard of public propriety that keeps us from thinking about sex in that very public context, but should there be? Is keeping sex out of public space actually creating the kind of sexual culture we want to live in? On the other hand, there are some implications of a system like this that are pretty scary – sexualizing public space is a double-edged sword in the real world of stalking, rape, and rampant misogyny.

Visible Identities
Relevant to: Identity and Awareness
Timeline: 5 years

Fashion and body language, while both ubiquitous and versatile, have their limits as a medium to surface identity. Not only that, but they feed things we do not always want to feed. Want to show the world that you opted out as a consumer? Great, there are product lines for that. The complicated details of a modern identity cannot always be easily read in person – see femme dyke invisibility, for instance. Turning our bodies into physical billboards only goes so far, can only express some identities, and is really quite inflexible – subtlety, the complexity of a person, gets lost.

Why bother trying to make a complex identity fit into the commoditized physical world? Splash it around your virtual self! Replace the image of you that people see with a mutable full-body avatar, and show the world an entirely different identity, without any of the complications of physical objects. Don't want to push things that far? How about a simple set of icons that can stand in for different facets of your identity and show off different slices of your data?

For anyone who has not been watching it explode over the past few years, augmented reality involves digital information, either in some rendered form or just simple text or flat graphics being inserted into a video stream shot in real time from the viewer's position. For example, one can hold up a phone to a scene, and have it act as a lens, appearing roughly transparent. The information overlaid is located positionally in the world, augmenting the view you have of the space.

AR has been exploding for a lot of reasons, and it seems pretty likely that it will continue to become an important component of how we interact with the world. Currently, short-range object locations and graphic registration are challenging for a lot of systems, especially lightweight ones – they are far better at overlaying data on building-scale objects and at map-scale distances. Registration and location will continue to improve as the technology matures. In order to be socially interesting, however, some degree of ubiquity is required, as with the previous fiction, at least among the class of people with whom you are interacting. Using AR to surface identity requires not only deep penetration of AR, but fairly regular, if not continuous use of it – your augmented identity will not be visible if no one looks. The notion that one might end up presenting completely different identities to different social classes is also interesting, if undesirable – the creation of a class of social have-nots, economically locked out from large swaths of not only culture but immediate personal presentation.

Avoiding Ambient Indoctrination
Relevant to: Awareness and Control
Timeline: 15 years

Advertising is, generally speaking, a very important source of cultural propagation in late capitalism. Specifically, advertising, as a core component of popular culture intended explicitly to cultivate desire, has a strong effect on things like body image and the sexualities that are visible to us.

Several years ago, São Paolo banned outdoor ads. Why wait for that to spread? It would take a fairly intrusive AR system, but in theory, you could literally have AdBlock for reality – imagine walking through a city without the visual assault of hyper-perfect heteronormative bodies, or riding mass transit without being told all the ways your body is inferior.

Of course, this goes both ways – what happens when the spammers (or worse, the "legitimate marketers") start to plaster themselves all over augmented realities? A potential escape becomes another front.

Making Sex Safe Again
Relevant to: Intimacy, Identity, Awareness, and Control
Timeline: 25 years

The sexual revolution at the end of the Modern period in the 60's and 70's was brought about, socially, by many things, and enabled by a whole host of others. Chief among the enablers was the Pill, which suddenly made sex less serious, divorcing it from reproduction. In the 80's and 90's, HIV and the rise of at least the awareness of rape culture made sex serious again, ushering in what might be called late heteronormativity. What if we could get rid of that seriousness again?

Picture yourself out drinking at a bar. There is a woman there with whom you have definite chemistry, but you just met. Still, you would both really like to get laid tonight, but you forgot to bring gloves. So, you both get out your phones. You might exchange numbers, but mostly you are trading a limited medical authorization. You press your phone against her abdomen and tap a button to wake up her implant, and then send a status query. The report comes back happily green on all the STI antibody tracking data, and your phone confirms the device signature. She checks you, and you head for the door.

Implantable medical devices are advancing rapidly these days. Currently, most implants are special-purpose devices focusing primarily on various kinds of therapeutic electrical stimulation. This is changing rapidly, though. Many devices are already capable of sensing a broad range of physiological parameters. Manufacturers have been working for some time on drug-dispensing implants. Implants which are intended for purely diagnostic purposes are starting to show up on the market, albeit slowly – implant manufacturers are very conservative.

In the research world, work is being done on communicating networks of implants. For instance, one might have one implant looking at impedance-based measurements of lung fluid, another dispensing an emergency bronchial dilator when needed, and a third archiving information and periodically talking to an external device via radio.

The challenges here are nontrivial, of course. Power management is a huge concern, as the devices currently run on batteries that must last five to eight years. Some newer implants use inductive charging, and there's a chance that low-power devices could run on biofuel cells that take power directly from the body, parasitically. The security and privacy concerns for inter-implant and implant-external communication are also nontrivial, especially as implants cannot currently afford the CPU or storage required to perform modern cryptography. Moore's law has a way of making at least some of these problems go away, however.

Advances in microfluidics, a technology wherein minute amounts of fluid are pumped around through channels cut into silicon wafers at a similar scale to the transistors on a chip, are allowing the rapid miniaturization of diagnostics that currently require full-scale laboratories. Disease detection is a significant research priority for medical microfluidics, and reusable devices that can detect multiple diseases are already on the research horizon. As the devices are constructed primarily via traditional semiconductor fabrication techniques, they're likely to follow fairly similar cost and capability trajectory.

As we come to understand implants better and the risk of implantation surgery drops, diagnostic implants are likely to become more attractive from a risk/benefit perspective. It is likely, even, that a general-purpose implant capable of continuously monitoring the state of the body and detecting a wide range of diseases accurately (and, given continuous monitoring and in-body positioning, much earlier than with conventional techniques) will be seen as a reasonable component of preventative care. As these devices become widespread, a flexible system of authentication and authorization will be necessary, if nothing else to handle cases of emergency medicine and child and elder care. Why not enable them for personal uses too?

This, of course, brings up all sorts of other scenarios. If someone reprograms their implant to lie about their HIV+ status and you are exposed, it seems likely that a crime has occurred. But what crime? The intentional HIV exposure or the reprogramming of the implant?

An implant could also be useful if you are out drinking – telling you when you are getting dangerously intoxicated, or, for instance, if it suddenly detects GHB in your bloodstream. While this might be very useful in making that night on the town safer, it could cut both ways. If you are a victim of date rape, could the rapist subpoena data from your implant to prove that you were not so intoxicated that you couldn't consent? Is an implant part of you, and thus subject to Fifth Amendment rights preventing you from testifying against yourself, in the case of something like a DUI charge? Or is it just another external computing device, despite its uniquely privileged location?

Orgasm Control
Relevant to: Intimacy, Identity, and Control
Timeline: Now

In some stereotypically masculine worldviews, orgasms define sex. If you came, you had sex. This blatantly ignores the fact that many women are either pre-orgasmic or will frequently not have an orgasm during many sex acts they participate in – even sometimes ones they quite enjoy. On top of this are all sorts of fetishes specifically about the act of having an orgasm, chief among them various forms of orgasm denial.

What if all this could be controlled at the push of a button?

Orgasms are a function of the same part of the nervous system that controls breathing, heartbeat, and similar activities. There is a point on the sacral nerve that can be used to trigger them with electrical stimulation. In addition to triggering orgasms, you can, with careful measurement, detect them, and by suppressing the firing of those nerves (by applying an inverted electrical impulse in real time), you can suppress them. The first of these is a function easily performed by current implanted medical devices – exactly that kind of neurostimulation is exactly what most of them already do, and successful clinical trials have already occurred in North Carolina. Sensing is somewhat more difficult, but entirely within our current reach, and while reliable suppression might be currently difficult, it is largely an engineering challenge, not one of fundamental research.

From a simple perspective, you could call this the ultimate sex toy. Push a button, have an orgasm. Dial knobs for intensity and duration. Decoupling orgasms from their normal context is interesting, too – how does that change the experience? As a dominance/submission device, what happens if you set up the control unit to work remotely and give the button to someone else semi-permanently, with an implant set up such that you cannot orgasm without it?

On the other hand, what about a monitoring-only device? How would a monitoring device change state treatment of sex offenders? What happens when you require a sex offender to account for the timing and location of every orgasm? While obviously the roots of sex offense are complicated, does a push-button orgasm help them re-integrate into society and resist offending again?

Other people might be interested in a monitoring device, too – it could be the ultimate purity ring, verifiable evidence that the person you are sleeping with has never had an orgasm, regardless of whether or not they have tried. And with push-button control, no more having to learn how to actually please someone sexually.

What would a device like this do to sex work? Would people with the implants still patronize sex workers, assuming they were otherwise inclined to? If so, how would it change that interaction? Would it become more explicitly about human touch, and less about getting off? Would sex workers use the implants themselves?

DNA Sequencing for Sex Workers
Relevant to: Control and Economics
Timeline: 5 years

Sex work occupies complicated, conflicted political space in the modern left. That said, it is safe to state that in most current legal regimes the majority of sex workers engaging in heteronormative sex acts are at a power disadvantage to their clients.

In order for a sex worker to operate, their clients must be able to find them in at least some context; the clients can often be relatively anonymous outside of the moment of transaction. The sex worker, already disadvantaged by this, is often more legally vulnerable while the client may only be socially vulnerable.

Imagine that you are a sex worker. Your most recent session has gone poorly, and the client did not pay. You are not in a position, normally, to ask for anything like identification that might give you any recourse. Your client, however, cannot avoid leaving DNA material behind. You swab a sample into a paperback-sized DNA sequencer. Ten minutes later, you anonymously upload the sequence and find a match in a public genetic database. The sequencer signs its sample with an anonymized but verifiable signature. You add a note describing the circumstance under which the sample was taken, and let Google do the rest.

DNA sequencers are getting amazingly fast and spectacularly cheap, and the trend shows no sign of slowing down. There are questions about how strong DNA evidence really is, but tactically, that may not matter – the social onus will probably still end up being on the person whose DNA was posted, to explain what happened or why someone wanted to frame them.

Outsourcing Judgment
Relevant to: Awareness and Control
Timeline: 10 years

People do a lot of stupid stuff while drunk, high, or otherwise not all that with it. A lot of that stupid stuff involves sex. Sometimes your friends will step in, but they're not always around, and they might be the problem. How much can someone be protected from themselves?

Take a pair of glasses with a camera in them that stream video remotely, and add a speakerphone and an implant that measures blood alcohol. Before you decide to go out drinking, you figure out what the limits you would like to set for your evening.

Later that night, you walk up to the bar to get another drink, and your glasses flash an icon at you, telling you that you have had enough. The bartender notices it and shakes his head at you.

Later, elsewhere, you have ended up somewhere secluded with someone you have been flirting with. A voice in your ear warns you that you are not sober enough to decide to sleep with someone, per the limits you set earlier. You ignore it. A little bit longer, and your phone trips over into speakerphone mode and informs your the other person that you are not sober enough to consent, and that per prior agreement, you are not currently legally in charge of your actions. You end up safely at home alone, without another embarrassing complication to deal with, along with a much less painful headache than you might have had.

Portable streaming video and microdisplays are already common enough technology, although combining them into a socially acceptable package has been a challenge for a long time. The implant technology is somewhat more complicated, but entirely within the realm of current possibility. The legal perspective for something like this becomes more interesting – is it possible to create a contract which gives away a temporary power of attorney to a call center employee who's acting on your prior instructions? While obviously the only meaningful solution for sexual assault is for people to stop assaulting others, would something like this fill a useful role in allowing people to balance inebriation and the morning after?

Pulse Bracelets
Relevant to: Intimacy
Timeline: Now

Long-distance relationships are hard. Keeping the kind of daily awareness and the feeling of connection going that one has with a local lover takes a lot of work. Technology has already radically changed how we interact with long distance relationships – Skype,

IM, SMS, and flat-rate phone plans have already made that kind of connection much easier to keep. Where else can it go?

The daily rhythm of life is expressed in many ways, big and small. Imagine sharing your heartbeat with a lover. You each have a bracelet that mirrors the other's pulse – a little membrane beating away, inside your wrist. Surprisingly simple things can carry a lot of information about your partner's life.

The technology behind this one is simple – a GSM modem, a battery, a heart rate sensor, and a small actuator.

Obviously, at first a bracelet like this would be distracting. Eventually, though, given enough time to get used to the stimulus, does it fall into sensory integration? If so, what does that end up meaning – can you tell when someone wakes up, what the rhythm of their day is, just from a heartbeat? How does it cross over into in-person interactions? There are some studies that have shown that even loud music can apply a synchronizing effect to the heart's rhythm – would this do that? What happens when you have multiple partners, and receive multiple heartbeats via different actuators?

Endocrine Management
Relevant to: Intimacy
Timeline: 20 years

A heartbeat is a single measure of physiological state, and a coarse one. What about the endocrine system? We are coming to understand just how heavily our mood, how our bodies perform, and how we think and perceive things are all dependent on a small number of brain chemicals. These are obviously very complex systems, to be tampered with very carefully. But, as we come to understand them better, we may be able to both monitor and alter the balances of these neurotransmitters in real time, again using implanted devices. While we currently act on the brain's chemical balance with drugs, these are, relatively speaking, very crude devices. In addition to allowing us to perform the same kinds of therapies much more accurately, working in real time allows us to introduce new behavior-oriented functions.

What effect could interacting directly with your partner's endocrine system have on the relationship? Imagine an endocrine mirror – carefully controlled, among other things to prevent feedback effects, but designed to nudge you toward similar affective states. While it could easily trigger entirely new kinds of codependency, it could also allow for radically more compatible interactions. Especially interesting would be the effect on distance – feeling what your partner feels, despite knowing that they are separated from you by hundreds or thousands of miles.

Even if the effects are isolated to a single person, situationally-based alterations of neurochemistry could have an interesting effect on relationships. Imagine if an implant gave you a steady drip of oxytocin whenever you interacted with your partner? Make that first month relationship honeymoon last forever.

Of course, if you are triggering neurochemical changes in response to outside behavior, you could trigger them in response to all sorts of things – similar to orgasms, you could give someone a neurochemical remote; an emotional organ of sorts, to play your brain.

Mutable Appearance
Relevant to: Intimacy, Identity, and Control
Timeline: 15 years

Virtual projections of identity are great, but what about actual physical changes? The catalog of the ways humans have altered themselves and their appearances is large and exhaustive, but the constant is that with the exception of things like hair and makeup, the body generally changes and is changed slowly – even clothing changes on the course of hours, unless you have an atypically hyperactive personal costuming department. What about creating situational physical identities? There are plenty of reasons to do this, and not just acceptance by a monoculture – fractured network social spaces provide for complicated terrain to navigate, and you might want to become several different versions of yourself in a day.

You wake up in the morning and go about your usual routine. You toss on a t-shirt and jeans and head for the train. On the train, you pull out your phone and dial your jeans and your t-shirt to the color that feels right for the day. You poke around online and find a cute little animation that looks good, and you grab a frame from it and toss it onto your t-shirt. Just a still – nothing too flashy for the morning.

You get to work, and the tattoos on your right arm fade in slowly; it's a relaxing day. Your skin changes a little through the day, fading blank when you have an older, conservative client to talk to, blending into a darker, more aggressive design as you bike across town in the afternoon.

In the evening, at the club, your shirt is cycling through something vaguely biomechanical, all gears and tentacles, and you see your recent ex-girlfriend. Your skin flashes black in shock for a second and then you slip out the back.

Products are already showing up on the market with flexible low-resolution liquid crystal-based wraps that can change color. Similar technologies are being used in experimental color-changing cloth. E-ink displays are improving in quality and falling in price – flexible and colored versions with faster update speeds are coming. While the cut of clothing may continue to be static for a while yet, color and graphics will become more mutable.

We may be able to make an implantable version of one of these displays as well, but there are other options, too. While I've largely looked at hard technologies here and not at bioengineering, one option that's too interesting to pass up is working with the color-changing chromatophores of cephalopods – instead of getting tattoos, get skin grafts. While the designs would be nominally under more mechanical control, might the basic biological nature of the grafts show through? The idea of humans having similar affective responses, changing color as a stress reaction, would be fascinating. Humans could literally grow entire new affective capabilities.

Soft-Body Tentacles
Relevant to: Intimacy and Identity
Timeline: 15 years

Profound alterations of the body are a longstanding fantasy for many people. They hold as much or more possibility of changing how we see each other, and were we draw the lines on what human means. Not to mention a whole lot of really hot sex.

Sadly, they are unlikely to become within the reach of our medical technology for a while yet. Even compared to the kinds of fairly invasive modifications we have talked about previously, large-scale body modifications appear to remain extremely difficult, if not impossible. Biomechanical hybrid systems, on the other hand, are potentially a different issue.

Would you give up an arm to have a prehensile tentacle in its place, assuming it could sense at least pressure and temperature?

Several emerging technologies are very promising when looking at biomechanically fictionalized bodies. The first involves functional electrical stimulation, a specific kind of electrical stimulation of the nerves. FES uses an external (or eventually, implanted) neuroprosthesis to stimulate nerves that have been injured. Current work has allowed paraplegic patients to stand and walk, and restored hand grasp function to quadriplegics. In addition, researchers are using the same technique experimentally to provide sense input from prosthetic limbs. Similar techniques are being used, again experimentally, to sense the activations of motor nerves at amputation sites, to drive the motions of powered prosthetic limbs.

Neuroplasticity is the phenomenon wherein the brain will reconstruct damaged portions and functions can migrate between radically different areas of the nervous system. It is implicated in all sorts of things, including stroke recovery and chronic pain. More importantly for us, it is also being used for modern prosthetics.

Using a combination of implanted sensing electrode grids and FES-style driven electrodes, a prosthesis can be attached to the ends of the nerves at an amputation site. Initially, the nerves will send and receive mostly noise, but via training on both the inputs and the output, a person can become attuned to the prosthesis. Their brain rewires itself as needed until they can use the prosthesis relatively transparently. One of the fascinating features of neuroplasticity, however, is that while it is helpful to use motor neurons for connections intended to move things, and sensing neurons for stimulation, it is perfectly possible, with training, to drive a prosthetic arm from a patch of nerve endings on the upper chest, for instance. This is sometimes useful therapeutically when the amputation site is not suitable for connecting a prosthesis and also interesting for our purposes. Prosthetics like this are on the cutting edge of what is currently possibly with today's technology, not anything that is in wide deployment. There are many challenges here, of course, including the number of nerve connections that a surgeon can make – even a dozen is fairly challenging, let alone hundreds or thousands.

Penetrating the skin is a big current problem for prosthetics, both for wiring up electrodes and for physical connections to mechanically anchor the limb. Creating structural connections to bones inside the body is, if not easy, then at least reasonably well understood – see joint replacement technology, for instance. In order to connect a prosthesis, however, there must be a mechanical connection to something that is by definition outside of the body. One of the most promising routes forward here involves gum tissue – cultured and transplanted onto the site of the attachment, it might provide that bridge.

This research is all aimed at traditional therapeutic implants, of course. Attitudes around prosthetics are a fascinating microcosm of how society treats disability and medicine in general. Research is almost always aimed at trying to bring a disabled person back to the social norm. However, once the basic integration with the human form is working well, the range of possibilities is huge. Aimee Mullins, a double-leg amputee who has done a lot of ground-breaking work in artificial legs, including legs which let her run at or above the peak efficiency of a so-called normal human, told a fascinating story during a TED talk. She has a dozen different pairs of legs that she wears for different occasions, including a pair that makes her several inches taller. The first time she wore them, to a party, a friend was shocked to see how tall she was, actually exclaiming how unfair it was – I mean, prosthetics are never supposed to make someone more able than "normal", right?

The prostheses she uses are static, not even powered, let alone wired into the nervous system. Additionally, the structural requirements for leg prostheses are more complicated than those for arms, as fewer shapes will accommodate walking. With arms, we can explore different forms, and we can even look at solutions entirely outside simulations of standard vertebrate bone-and-muscle movement.

Another interesting area of research emerging research involves what is called soft body robotics – basically, things that move by the same mechanisms as slugs, worms, caterpillars, or jellyfish. Instead of having a rigid skeleton with rigid joints that are actuated, soft volumes of tissue-like things change shape and volume. For example, one current Department of Defense research project involves a surveillance robot that can move like an earthworm and can change its shape, so it can squeeze under a thin gap under a door or a through narrow pipe.

Fabricating soft-body robots is, needless to say, very complicated. However, there's a lot of promise in 3D printing technologies, especially multi-material printing techniques.

Combining these different technologies, we see a fairly realistic path to construct a functional tentacle, either as a replacement limb, or possibly even just as an additional limb. The process won't be either fast or easy, but there's no reason to believe it's not possible, and not that far off.

How will people react, the first time they see a person with a tentacle for an arm walking down the street? How will they react the first time they end up in bed? If you lost an arm, what would you want to replace it with?

Emergency Bootycall System
Relevant to: Intimacy, Awareness, and Control
Timeline: Now

Fuck buddies are great. Everyone should have at least a half dozen. Scheduling, though, royally sucks.

You decide you'd like to get laid tonight, but you don't really have anyone specific in mind. So you send out an SMS to the Emergency Bootycall System. It propagates out, first going to your A-list, and then if no one has taken you up on it, your B-list and your C-list. You've told the system that you'd like dinner beforehand as well, so the system looks at the preferences you and the respondent have both set and selects a location. It looks at your calendars, figures out when you're both going to be off work, and sets a time, and if necessary, sends a reservation request.

Say everyone you already know is busy. The system asks you if you like to throw out a wider net. Maybe you'd like a surprise. You tell the system to set up a date. It tells you to be at a specific place and time, and you meet someone there. You've already told it what you're looking for, in general and tonight, and it knows who's off limits. The people whom you've slept with have vouched for you, in a distributed reputation system, and other people have decided how much they trust those people's opinions, and about what, so when a stranger sees your request, they have a measure of what the community thinks of you, and how distant you are socially.

Enabling Nontraditional Reproductive Groups
Relevant to: Intimacy, Identity, Control, and Economics
Timeline: Now

This concept and the next are less design-related than most of the rest, but they have implications that are too interesting to pass up.

Given the biological complication of a pregnancy, it seems unlikely that we'll see reproduction literally and completely divorced from sex any time soon. That said, gene manipulation has already allowed us to create embryos with more than one genetic mother, more than one genetic father, etc. This is obviously a good start toward a non-traditional reproductive family unit.

Breastfeeding is a fascinating and complex interaction with a new child; we're still learning all of the things it does. Among other things, it appears to encourage an emotional bond above and beyond the intimacy of the experience. There's also a transmission of disease immunity via breast milk. Traditionally, of course, breastfeeding only occurs with the birth mother, not other members of the family (excepting wet nurses). This is limiting even in traditional heteronormative families, as the father is excluded from this relationship.

Medically, though, there's no reason for this – inducing lactation is medically trivial, for both men and women. Why not use this ability to redesign the non-traditional family unit? Criticisms are leveled at non-traditional families raising children for not having a natural connection to the child. Obviously, these are more politically motivated than they are motivated by any meaningfully objective understanding, but why not redesign that relationship further, to strengthen that natural connection? As it is, breastfeeding is frequently complicated for mothers who want to work, acting as another site enforcing the heteronormative economic relationship of motherhood. If all of the care-giving adults in the family can breastfeed a new child, that problem largely goes away. Of course, the existing construction of masculinity would be challenged by this, but is that a bad thing?

Nontraditional Group Legal Structures
Relevant to: Identity, Control, and Economics
Timeline: Now

Gay marriage is fine and all, but really, is that even what we want? Congratulations, the heteronormative relationship contract has been extended one tiny notch. Why not fight to blow it out of the water, instead?

In the 1886 decision in the Santa Clara County vs. Southern Pacific Railroad case, corporations were granted rights as natural persons, with almost no precedent. Throughout their history, corporations have proved to be fascinatingly flexible legal entities. Why not use that same structural construct and design a modular structure of contracts to enable a whole spectrum of legal relationships between whatever groups of people decide they want a legal relationship? Of course, designing a set of contracts and ensuring that they'll be legally enforceable are different matters, but without an existing model, there's nothing to fight for.

Family Cohousing
Relevant to: Identity, Control, and Economics
Timeline: Now

The economic structures of heteronormative relationships reinforce those relationships in the culture to an impressive degree. If we want to break the pattern of relationship structures, we have to break the economic relations that create, support, and enforce them. Divorcing economic structures from reproduction is just as fundamental of a social change as divorcing sex from reproduction is. This, of course, is an amazingly complex problem, many parts of which have more to do with social norms and economics than design. One area that is intimately connected to design, however, is the way those heteronormative economic reactions are programmed into and enforced by the built environment, the architecture around us, the houses we live in.

Raising children is hard. Some form of income and labor sharing is hard to do without, and it strains the resources even of many couples. The dominance of the intimate couple as the core of the family unit is reflected in housing, and attempts to construct alternate sharing arrangements literally don't fit.

Picture a group of three to five people, all either raising their own children or committed to help raise the children of other people in the group, but none of them involved with each other, although they may have partners, even serious long-term ones, outside of the group. Where do they live?

One model would be a cluster of small apartments, each with a bedroom each for the parent and the child or children, a separate bathroom and a small private living area. All of the apartments would share a common kitchen and living area. Trying to shoehorn that relationship structure into a set of more traditional apartments might work, technically, but it's very difficult to create that sort of social structure without support from the environment.

Even if the group can afford to have custom construction work done, it is difficult or impossible to build housing like that, between building code requirements and banks that won't lend money for non-traditional structures. Needless to say, economically disadvantaged parents, who could benefit even more from a collective structure, are even less likely to be able to find or create such housing.

There are almost certainly a wide variety of housing types which should be investigated, along with the corresponding relationship structures that they'd support. Until we can get variety into the available housing stock, heteronormative relationship structures will retain a huge amount of power.

Conclusion

This paper covers a lot of ground. I hope that some of it will give you new ways to look at how we do and can construct relationships, have sex, and live our lives. All of these things are mutable, and redesigning the context in live can change the way we live.

Also, remember – sleep with your local designer! The future of sex depends on it.

Eleanor Saitta
Structure Light Design Research Collective
ella@sldrc.com
@dymaxion

R.U. Sirius

HOW SINGULARITARIANS VIEW SEX

Note: this is written in a conversational style and is mostly made up of quotes from other people, including quotes from the article Sex & the Singularity, *published in the Fall 2009 edition of* h+ *magazine.*

Right now, the Singularity 2009 conference is happening in New York City. A quick scan of the lectures that are taking place finds no mention of sex, eroticism, play, fun, or feeling good. There is one lecture that includes humor and creativity in the title. So to paraphrase Paul Kantner of Jefferson Airplane from 1970, it might be necessary for those who believe eroticism is important or even central to a worthwhile life to hijack the singularity.

I called this talk how singularitarians view sex. But I really meant how singularitarians and other transhumanists imagine sex *after* the singularity or after some sort of transformative future. Probably, most of them view sex today more or less like most other people... or at least like most other geeks.

But I can think of at least one exception. I had an interesting conversation on the way to a transhumanist meetup with a fellow who told me he doesn't like sex. He finds it an annoyance and a distraction from thinking about and working on more interesting things. Now, this is a good looking young guy with an attractive wife. But I found that very honest. I do suspect that more singularitarians would like to get rid of sex than would be inclined to talk about it. But not a majority.

In the '60s and '70s, we thought that war was made by old men who were alienated from their sexuality – men like Robert McNamara and Dean Rusk seemed like disembodied bureaucrats. And as youths, if we kept on having good sex and if we learned to move our alienated white asses to rock and roll, we were not going to become bureaucratic warmongers.

Of course, that was simplistic. Think of Ted Nugent or Kid Rock. We're ruled by generations that have rocked, and we still make war.

I organized an opinion poll for the current edition of *h+* magazine. As I wrote in the intro: We asked several leading thinkers in the radical tech community the following question: Is there sex in the posthuman or singularitarian future.

Nobody blatantly renounced sex as a low-level animalistic bore, but I would start by sharing some comments that align somewhat with that view.

Theme 1: The end of sex, or the de-emphasis of sex

Natasha Vita-More is one of the founders of transhumanism as an activist identity, starting back in the 1980s: She opens her response in the magazine with this: Will there be sex in a posthuman future? Yes. It will not look, sound, or even feel like the traditional act of rubbing membranes against each other. The aesthetics of posthuman sex takes a giant leap into unchartered territory. If the posthuman is semi-biological, then the physiology of sex will remain the same sexy, smelly, wet sex that we so dearly love, but with added twists in virtuality and simulations. But what happens when our human genitalia is gone? What will we rub instead? (the answer is coming later)

Michael Anissimov is a popular transhumanist blogger who is deeply involved in the Singularity Summit and a regular contributor to *h+* magazine. He said: "Sex is quite a simple act itself – much simpler cognitively than me writing this paragraph. Sex has existed for hundreds of millions of years, but general intelligence has only existed for a few hundred thousand. Sure, sex activates higher cognitive functions, but that is a credit to those functions, not sex itself. It is important to remember that (emphasis mine) *sexual intercourse is a highly ancient, simplistic-at-its-core activity that we may choose to discard at some point in the future in favor of more complex activities that generate even more pleasure and connection between people.* Whether we choose to call it "sex" will be entirely arbitrary, but it may bear little resemblance to the sex of today.

So, to paraphrase Cole Porter, birds do it, bees do it... so *really*, it's not that interesting.

Ben Goertzel is the CEO of the AI companies Novamente and Biomind. He's our *h+* AI columnist and a leading figure in singularitarian circles. In his response in the magazine he is even more explicit about the end of sex as we know it. He said: The experience of gaining pleasure via in some sense merging with another being... that will probably survive the Singularity, but will likely be customizable into various forms, which may end up bearing little resemblance to "sex" as we know it today....

Athena Andreadis is a biologist who is very critical of the singularity and some of the more utopian predictions in transhumanist circles. This is a quote, again from the magazine: Jacking Off while Jacked In... cryonics, robotics, uploading, singularity by AI... the concepts speak for themselves: no Eros, only Thanatos – at best, endless (and simulated, yet!) masturbation in VR lotusland. Besides, if you're obsessed with control over all your functions, how are you going to let go enough to have an orgasm?

Theme 2: The end of sex for procreation

In the same article, the famous singularitarian Ray Kurzweil points out: We've already separated at least some of the original biological function of sex from its social and sensual function...

Karen Elliot from the *h+* Facebook contributors group offered this comment: "You should talk about the end of sex as the engine of evolution – we're fast approaching a point in time where we'll have control over our evolutionary futures. Sex will no longer be the ipso facto engine of evolutionary change. This has *huge* implications for how we, as a society, operate. Also, think about the consequences of deliberately gaming sexual selection mechanics. Think of the pick-up artist community, spearheaded by evolutionary theorists like Erik Von Markovich – the subject of the book *The Game* by Neil Strauss.

There's a community of men (and women) who are studying psychology and sociology to better understand how to seduce people. What happens when attraction becomes science?

Theme 3: Simulation... Subtheme: Multiple selves

Ray Kurzweil says: We will continue to have bodies in the singularitarian future, except that we won't be limited to just one. We'll have different virtual bodies in different virtual reality worlds, and morphable, nanobot swarms for real worlds. A couple could become each other in a virtual reality environment and experience the relationship from the other's perspective. We'll be limited only by our imaginations. That will be true in general for virtual reality, which is where we will ultimately spend most of our time.

Returning for more of Natasha Vita-More's response, she said: Exosex, sex outside the biological body, would be simulated in virtuality, much like Second Life or Skype, and other digital formats where sex is enhanced, extended, digitized, and synthetic. It would be more real than real – a hyper-real experience. *Endosex,* sex *within* the body or form, would exist even if the posthuman is so-called disembodied or, better, a distributed collection of selves (multiselves) co-existing on multiple platforms, including biological personas, virtual avatar personas, or other types of forms in different substrates and platforms.

Extropia de Silva is a notable Second Life figure. This is from her response in the magazine: "Polyamorism will be the norm. After all if 'I' have uploaded, duplicated myself and exist as self-similar copies in cyberspaces co-existent with realspace, where does the 'self' end and the 'other' begin? Relationships will be tried out in simulation, combining variations of each self, weeding out combinations that do not optimize cooperation & mutual gain. Selective memory editing may be used to erase memories of sub-optimal relationships, leading to love affairs that are always subjectively ideal.... It is interesting to note that humans rather enjoy romantic period dramas. For instance, Jane Austin's books concerning the trials and travails of love in upper-class society remain as popular in the 21st century as they were when first published in the early 1800s. If posthumans inherit their predecessors' love of historical romances, they might simulate the relationships of ancestors in the dim and distance past. In Fact, given the vast computational resources that Moravec, Seth Lloyd and Nick Bostrom have appealed to, it is perhaps astronomically more likely that, if you are in a romantic relationship right now, it is one being simulated by godlike intelligences, rather than being real in a physical sense.

Theme 4: Intensification, pleasure, aplification, issues around pleasure

Alex Lightman is the Executive Director of Humanity Plus, an organization not directly related to *h+* magazine. In the magazine article, he responds: The primary purpose of the Singularity will be seen, after the fact, to be Awesome Sex. There will be exponentially *more* sex, with exponentially more interfaces, and with exponentially more measures of pleasure. First, whole brain emulation (a more stealthy way to say uploading), will enable us to make almost perfect replicas of our brains, which can then imagine, aided by the cloud that puts thousands of supercomputer-equivalents at our beck and call to generate millions of sexual fantasies and to engage in variations of them. There will be no limit to the number of our own brain replicas we can create, host, & send off to have great sexual adventures in imagination, & then bring the "best of" back for reintegration. Second, we will be installing bioports into our body, a la *The Matrix* or *Sleep Dealer,* each of which can stimulate our nervous system. In heterosex, men penetrate women, but with this, men and women will interpenetrate each other multiplee, and, as with USB 2.0 daisy-chaining, so will men, women, & androids be able to multiplee-interpenetrate, locally or remotely.

Back to Natasha Vita-More: Sex is all about nerves, and the human brain is the pleasure center. Nerves would be replaced with synthetic fibers or electrical charges that continue to detect & transmit nerve-like sensations. The posthuman's post-neurobiological brain would experience a series of sudden spasm, contraction, and surges, a sense of pleasure and release. *For the most part, we could be rubbing neurons* (emphasis mine). If we are distributed multiple-selves, co-existing on different platforms, a sexual experience could be a community event or a selection process to determine how many selves would be involved. The entire field of posthuman sex could give new meaning to sex freedom & gender differentiality – where a person could have different scenarios, depending on what form or type he/she is in. Human form: membrane, wet sex. Semi-human form: neurological ecstasy. Post-human form: multiple exchanges of digitized codes reaching a crescendo. But sex is not just about the crescendo. The physical and/or electrical charges brought about by excitation could be relocated to different parts of the brain. In a bio-body, instead of reaching climax, a person could have that energy-charge redistributed to the memory center of the brain for deeper focus. Likewise, a non-bio-body could use the energized charge for a totally different activity.

Michael Ames, from the *h+* contributors Facebook group, raises the theme of controlling and amplifying pleasures automatically. He writes: Pre-singularity, my view of sex is: Sex is a genetically programmed drive/impulse/act that I cannot ignore/change/adjust-the-volume-of, so it should be incredibly annoying, but mostly isn't because it is so much fun. Post-singularity it will be way more fun, because I get to "twiddle the knobs" at will, knobs labeled Volume, Gender, Stamina, Love, et cetera.

Returning to Michael Anissimov, in the magazine – he sounds a similar theme. "The connection between certain activities and the sensation of pleasure lies entirely in our cognitive architecture, which we will eventually manipulate at will. It's probably less complex than we think – many drugs can directly stimulate the pleasure center, and these are much simpler than brain-to-computer interfaces. With sufficient ability to intervene in my own neurology, I could make any experience in the world highly pleasurable or highly displeasurable. I could make sex suck and staring at paint drying the greatest thing ever. It scares some people to think that the connection between pleasure & experiences is entirely arbitrary and not based on some deeper philosophical meaning, but too bad. We will likely choose to preserve sex as a highly pleasurable activity, but perhaps other people will decide to elevate philosophical discovery or artistic creativity to a higher pleasure level than sex. That's entirely their decision.

Theme 5: Borgian sex & brain links: come together

Extropia de Silva: A committed relationship would be to accept a complete merging of two selves. True love would be expressed by transferring the two uploads into a single, higher capacity "brain" (such as the sentient Internet itself) in which both minds run simultaneously. Such "twindividuals" might merge with others, resulting in an expanding hivemind. Parts of the brain could be distributed over large distances, though if communication delays cannot be overcome that would impose a limit on how far the society of selves can expand and still be ALL=ONE. Possibly, group-minds that expand far enough to experience significant communications delays will fragment. These, as well as others initially seeded from other *twindividuals* might expand until they are bounded on every side by neighboring group-minds. Moravec has speculated that competition for space, matter, and ideas might result in "vile offspring" (Charles Stross' term for posthumans that have diverged from the human species to the extent that nothing recognizably human remains within them) devouring the physical substrates of neighboring group-minds, "space, energy, material and useful thoughts reorganized to serve another's goals."

Repeating a quote from Ray Kurzweil, which also fits this particular theme: "A couple could become each other in a virtual reality environment & experience the relationship from the other's perspective."

Alex Lightman: One of the most profound pleasures of sex, in my experience, is what I call the "empathy hall of mirrors effect." That is, to be able to not only feel what you are feeling, but also feel what your lover is feeling, almost as deeply as she is. If she is also an empath, then you can get a positive feedback loop of sensation. After the Singularity, most transhumanists who choose to stay embodied will *present* as empathic metamorphs, possibly surrounded by utility fog that enables us to become anyone, or anything, seemingly anywhere, and, with telepathy common, to be able to transform ourselves into our lover's heart's desire at a moment's notice.

A member of the h+ contributors group and writer for *Wired* who asked to remain anonymous wrote: "Imagine a future where direct brain-to-brain hookup is a common part of sex. Will a b-t-b during sex reveal to you that your wife has been fantasizing about Brad Pitt all these years? Or Homer Simpson? Or her father? Or YOUR father? Will you want to know that? And do you dare let her know what is going through *your* mind during sex? I wonder if direct brain-to-brain hookups are going to be a *huge* generation gap when it comes to sex. No matter how open to new experiences you are, as you age your brain becomes less and less plastic. Also, we hope, you also become somewhat wiser, or at least more mature. How will that work when you're in your 40s and want to jack into your favorite 19 year old? Speaking for myself, I might physically desire a 19 year old, but I'm already at the point where a conversation with most 19 year olds is more effort than it is worth. I can practically assure you that I don't want to see what is going on inside a 19 year old's head! Will most people want to visit the cacophony of the teen brain when they're just trying to get their rocks off?"

Darren Reynolds of the *h+* Facebook group commented: You could connect your neurons to mine and we'd become a single conscious entity. That's closer to the ideal of holy union than anything current organisms achieve. Perhaps eventually we will leave these human bodies to travel through space. Perhaps clouds of dust, intelligent and conscious on a scale beyond human imagination would encounter each other as they expand through the universe. At first low bandwidths are used to explore each other, with the rate of information exchange growing to a climactic explosion until the beings effectively become one. Ultimately we'll all end up the same conscious entity, using all energy and materials in the universe to maximise the amount of consciousness. I see that as the inevitable outcome of life, whatever laughable messing about we humans do in our silly parliaments.

Natasha Vita-More: Sex is a means of communication. In posthuman futures, connectivity is paramount. All the connectivity – from simulated environments to the noosphere could end up being one very big bang.

For contrast, here's an extreme case of auto-eroticism from CJ Carr, responding from the *h+* Facebook group
I think it'd be interesting to be able to have sex with yourself. You could do it in virtual reality, controlling two bodies at once. It wouldn't be too difficult if you did some sort of mirror-image movement thing. Seems like it would work, but you could definitely take it way beyond that.

Theme 6: The cosmic imaginative

Adam James Davis, a respondent from the *h+* Facebook contributors group said: "A concept I like to call cosmodildonics: Ala the mechanical vibrations of an electric dildo, erotic stimulation from any and/or all repetitive and periodic phenomena in the universe from electromagnetic waves to the orbits of planets; from the oscillations of civilisations rising and falling to the potential big bang/big crunch/big bang etc. patterns of whole universes. I've garnered euphoric and intense experiences through the contemplation of seemingly inescapable thought loops, contradictions and paradoxes... they have built up and built up. I believe this could be a system for posthuman erotica... or the feeding of information into a closed system preventing such information's escape, building up until the closed system explodes (orgasm), but with *abstracts* and *concepts*. The conceptual, abstract equivalent of this idea... to program oneself to oscillate rapidly between extremes at the ends of philosophical and ideological spectra a la the peaks and troughs of mechanical vibrations of mechanical dildos. And then there's Dadaist transhumanism: To take the Dadaist/Surrealist principle "Everything is Erotic, Everywhere Erotic," to a literal extreme through technology... to program oneself in order for everything and everywhere to induce erotic stimulation and orgasm, including the thought of de-activating such programming
to get on with life... the antithesis would be: "Nothing is Erotic, Nowhere Erotic" *(R.U. Sirius comment: Like Being at most tech conferences.)*
Postposthumanity: Infinite members of infinite genders, in an infinite orgy... and eternal, megascale ejaculations. Orgasms through supernovae! A car crash is erotic... colliding celestial bodies are erotic.

David Jay Lewis, from the *h+* Facebook users list said: "I predict, we as Human Beings, will realize our symbiotic nature and proceed to create new forms of symbiosis with the animal kingdom. Sex will ultimately become not merely a state of mind but mind itself. How will those in the generations leading up to the Singularity view sex? A libidinal adventure that turns us from mere desiring machines to active participants in an erotic paradise that is closer to the visions of the ancient Greeks than is to our current industrial images of decay. A world in which it will be possible to engage any of the Goddesses as Pan, or to seek pleasure as a plant might experience the honey bee. A nanoforeskin could literally filter out the HIV virus and overnight turn our planet from one in which there is a plague, to one in which sex is something we do for pleasure not simply out of necessity.

Final thoughts

Most sexual activity takes place in the mind and in the realm of the imagination. In asking about post-singularity or posthuman sex, you invite fantasy about fantasy or even something more recursive. My exploration of ideas about sex in the Singularitarian and Transhumanist communities reveals some diversity in implicit goals, from the replacement of sexuality as we understand it (but pointedly, no one suggested eliminating pleasure), to the idea that really great sex is the ultimate goal. There is perhaps an underlying distress at the messiness and uncontrollability of sex in current human form, and at least one person doubts that postsingularitarians could enjoy sex because messiness and being out of control is sort of the point.

Annalee Newitz

THREE SCENARIOS FOR THE FUTURE OF ROMANTIC LOVE

We all know the future of sex involves robots and teledildonics, but what will love be like in centuries to come? Here are three possibilities, based on current trends.

Serial and Parallel Monogamy

What it might look like:
A discovers sex and love at roughly the same time, among his group of friends. Some of them he's met at school; others are people he knows from social networks. He and his friends don't think of hooking up and dating and being friends as different things – it's hard to say where one ends and the other begins. As a result, A has sex which is as casual as meeting up over coffee, and friendships that are as intense as first love. And vice versa.

When he grows up and starts to think about settling down and having a family, his models for love and intimacy are based on what he experienced when he was younger. He considers love, friendship, and sex to be overlapping and interchangeable. For several years, he lives with three close friends. He has sex with two of them some of the time, and eventually one of them decides to get pregnant. The two of them decide to become a monogamous couple to raise the child, but remain living with their two other friends.

Years pass, and A and the mother of his child both fall in love with other friends. They decide to form a poly marriage, where they remain a couple but also have other long-term relationships too. Their two housemates have sex with each other once in a while, but start fighting. One of them moves out, and a new friend moves in. A winds up having sex with the new housemate one night when his two long-term lovers are off vacationing with their kids.

Where will this scenario come from?
In the west, changing norms around marriage have already made serial monogamy a reality for many people. They may be monogamous, but they will have several partners throughout their lives. Add to this the changing ideas about friendship and sex that is popularly associated with the social network generation, and you have a population of people who expect multiple partners drawn from extremely interconnected but casual friendship networks. As a result, long-term romantic relationships start to look more like friendships. The emotions are no less intense, but the structure of the relationships might take on the characteristics of friendships today: Constantly-changing groups of people whose feelings for each other range from talk-every-day closeness to casual meetups at the pub. Stability will be provided by the network, and by a few long-term close connections like A's monogamous relationship and later his two long-term lovers.

The idea that humans will one day live in poly marriages is a popular one in science fiction, and can be found in novels by authors from Ursula Le Guin and Charles Stross, to Octavia Butler, Iain M. Banks, and Robert Heinlein.

The Female Minority

What it might look like:
B is always one of four girls in her classes at school. Everybody else is a boy. At first this seems normal, but then they all go through puberty and she starts to realize that she is the focus of intense attention from all those hormone-charged boys. In the country where she lives, girls are considered less valuable than boys, but you'd never know it based on how the young men treat her. In fact, B manages to grow up believing that girls are more special than boys, because after all she and other young women are the objects of fascination and desire among their peers wherever they go.

In college, B falls in love for the first time after going on hundreds of dates and being told no less than two dozen times that she's broken some young man's heart. She's received gifts and plaintive love letters and weird homages but all of it made her feel weird and slightly guilty until at last she meets a man who shares her passion for puzzle games.

Of course, it's so hard for her to know what men are thinking. That's why B's romance almost didn't bloom. On their first date, she tells him all about her favorite kinds of word puzzles and her college classes and where she grew up and he just nods and smiles like all the other young man did. Occasionally she can pry some detail out of him about himself, but half the time when he's talking about himself he's really talking about his family or demurring that her opinions on most things are probably more informed than his. Finally, though, she challenges him to a game of chess and sees that they actually do have something in common.

Years later, he admits that he waited by the phone almost constantly waiting for her to call about a second date. She was busy with exams and didn't manage to get back to him for five days. While she finds having a steady boyfriend a relief – at least she isn't pressured by all the other men anymore – her female roommates in college are enjoying playing the field. They go to meetups and matchup balls and speed date events, amused by all the ways the men get gussied up and try to grab their attention. Her friends explain with bursts of giggles that some men prefer each other's company to these awkward competitions, and there are bars and clubs where no woman ever goes – she only hears rumors of them.

Where would this scenario come from?
In many parts of China, medical technology has merged with traditional beliefs and population control to leave some regions of the country with 150 men to every 100 women. This imbalance was produced after just one generation, and we may see repeats of this scenario in other nations where governments try to limit the birthrate. Many people still regard sons as the only way to continue the family line and ensure that elders will be cared for by a stable breadwinner.

The result of a skewed gender ratio, however, may wind up reversing gender roles. Men who want to get married will find themselves in the position that marriage-minded women were once in: Waiting by the phone, trying to please their dates by not speaking up too much and seeming too opinionated.

Ian McDonald has written about this in his short story collection about a future India, called *Cyberabad Days*. A twist on this scenario appears in Margaret Atwood's *The Handmaid's Tale*, where women are plentiful but fertile women are not. Women in Atwood's novel become the property of men, and there is no gender role-reversal.

Neo-Courtly Love

What it might look like:
C and D are raised in an affluent community where everyone goes to church, elders are respected, and a King rules the land. From a young age, C and D have known that they are promised to each other as husband and wife – it was arranged by the priest and the community's prominent families before they were born. All marriages are arranged, except among the poor, and C and D have only seen the favelas from a distance when they pass through the community gates in a tram to travel to a neighboring town or air station.

Nobody expects C and D to love each other, least of all C and D. They will certainly make a home together, raise children, and take care of each others' parents when the time comes. But they will seek love outside the bounds of marriage. They call it courtly love, after the medieval notion that marriage is for duty and romance happens in highly codified, covert ways that everybody knows about but politely pretends not to.

C and D are married when they turn 16, and their families buy them a small starter house in the heart of town, near the shopping mall. C works in her mother's hat shop and D is going to school to become a biotech entrepreneur like his father. Although C and D like each other, they cannot imagine a romantic spark growing between them. Passion has no place in an orderly home.

And so they both discover love a few years after they are married. C meets an intense young man from outside the community who aspires to one day own his own home. He works in the mall as a physician's apprentice, and C's effortless, money-bought beauty embodies everything he hopes to have for himself one day. He sends her secret poems via an encrypted channel; they meet in places nobody will ever find them. D knows she has a lover, but as long as it never interferes with family dinners and she leaves no clues anywhere, he is happy. D has a lover too, a waitress who works at the gentlemen's club where he goes with his father. She always serves him port with a salacious smile, and his liaison with her is looked upon as the sweet folly of a young man.

Where would this scenario come from?
Courtly love, historically, grew out of a society infused with traditions that were so old they felt more like window-dressings than true social mores. It was also associated with the ultra-rich aristocracy, who had time to engage in court intrigue and romance rather than having to work for a living. We can see certain trends like this in our world today, where ancient, devout societies pay lip service to tradition while indulging in decidedly modern activities with a wink.

As strong religious cultures from the Middle East slowly blend with the secular and religious cultures of the West, it's possible we might see the emergence of a neo-courtly love tradition. Especially among the wealthy elites. People who value the old ways, but want to experience Hollywood-style romance, may find themselves in a marriage very much like C and D's.

Authors like Robert Charles Wilson (*Julian Comstock*) and Elizabeth Bear (*Carnival*), who have written about neo-traditional cultures, often touch on this idea of people who lead hidden, unconventional lives in conventional society. Steampunk novels and Neal Stephenson's *The Diamond Age* have a touch of neo-courtly love in them, as do many modern fantasy novels like Jacqueline Carey's Kushiel's Legacy series.

Katrien Jacobs

PROBING ADULTFRIENDFINDER.COM
SEX AND LOGIC WITH THE HAPPY DICTATOR

Introduction

This paper investigates a diverse group of web users in Hong Kong and their sexual behaviors and pornographic self-representations as observed on the sex and dating site http://www.adultfriendfinder.com (hereafter "AFF.com"). The paper is based on an extensive case study carried out over two years, in which I functioned as participant observant and interviewer, or "coaxer", of selected AFF.com members. Their words and actions will be quoted, though they wished to remain anonymous for this study. They shared their sex experiences by means of online communication, online story-telling, and face-to-face encounters, some of which were recorded on audio tape or video tape.

Besides analyzing these dialogues with members, the paper looks at a sample of member profiles to discuss self-representations as "impression management," or the way web users manipulate and tweak texts and images to make good impressions on others. The aim is to observe impression management in relation to the site's unrelenting promotional campaigns suggesting sexual conquest through technological competency. The profiles were sent to the author's email account over a two-month period, more particularly between July 15 and November 15, 2008.

AFF.com is one of the booming web sites of a lurking "porno 2.0 revolution," a new trend in pornography and social networking described by a German art organization this way:

> "Users can upload movies and pictures via video transmission and web portals, creating social networks and digital parallel worlds. (…)As everyone can put their own content on 2.0 websites, the boundaries between porn producers and consumers become blurred. Users create digital doubles; biological constraints of changeability do not apply to them. Perfect beauty can be programmed." [1]

The AFF site is primarily a site for heterosexual sex and dating activities, where people upload their own home-made sexy and/or pornographic content to seduce others into sex encounters. The site is owned by the corporate network, Friendfinder Inc., which merged with the giant of sexual entertainment, Penthouse, in December 2007. It then became an ever expanding "family" of sites that reaches masses of people in different cultural habitats, all of whom are trying out the "porno 2.0 revolution" as a new type of sexuality and pornographic identity.

Friendfinder Inc. uses the same generic information architecture and web design for all sites within the family, and there is little or no concern for cultural differences. The paper tries to uncover the scripts of sexual behavior and the ways in which people in Hong Kong are interpreting them. Are they discovering new freedoms and novel sexualities, or are they irritated and hampered by the USA-style rituals and homogenizing sex scripts? The AFF zone will unfold as a case-study in pornographic self-representations from a critical angle, but without reverting to gridlocked moral debates about pornography in either the Western and Chinese tradition. Hong Kong is positioned as a unique material-historical place within Chinese culture, but is also further analyzed as a dot within a transient Internet-based network.

Aff.com as the Happy Dictator of Sex Affairs

Internet sex can be seen the epitome of a new type of "media intoxication," as predicted by early media theorists Walter Benjamin and Guy Debord. Web users are happily exploring daily activities and friendship networks like Facebook and MySpace, while also using the Internet to arrange sexual affairs. In the recent trends of Internet sex and porno 2.0, people are using social networks to feel sexual sensations and to retain memories. Benjamin's philosophy predicted a loss of feelings of "authenticity" or "satisfaction" in such mediated encounters [2], Guy Debord defined our craving for sensations as one of internalizing spectacles, as our entire relations with other people would become dominated by images that we have already seen on television, in magazines, or most recently on the Internet. Debord predicted that the modern individual would be intertwined within media spectacles that would hold so much sway as the ruling economic class would be perfectly portrayed by it.[3] These philosophies have captured a spiritual exhaustion and alongside modes of euphoric media consumption in capitalist societies.

AFF.com users are driven towards blissful networking and digital subjectivities, yet their actions and discourses also reflect the racial divisions and gender inequalities that are typical of 21st Hong Kong culture. Who are the winners and losers of the Internet sex conquest? In conjuring up the economic foundations of Friendfinder Inc., I reference the work of media activists Alessandro Ludovico and Hans Bernhard, who used the notion of a "Happy Dictator" to question the monopoly of the giant of e-commerce, Google, in their project Google Will Eat Itself (GWEI) as they write:

> How can a dictator be funny for the people? One chance is to know how to entertain people, while continuing to influence every decision they make, so invisibly maintaining the totalitarian power untouched. Google's management knows very well how to entertain surfers…On the Google planet everything works and is funny. Everything is light (as the interface) and tasty (as the images search), resource-rich (as Gmail) or fast and updated (as Google News)[4].

Just like Google, Friendfinder Inc. tries to be the funny dictator of Internet sex, even though sex in itself is a life-style and perhaps not even as vital to contemporary web users as the Google search engine. Nor has Friendfinder Inc., managed to launch those very slick and tasty applications that have seduced the masses on Google or Facebook. But Friendfinder Inc. is a crude business venture that so far has mastered the game of social networking, while primarily making revenue through its membership programs along with Internet advertising campaigns.

AFF.com Fact Sheet: Come Fuck Me or Perish

Friendfinder Inc. was founded in 1996 by a Silicon Valley company called Various that pioneered different sites for sex and dating. In December 2007, the site made news through a ground-breaking 500 million dollar sale to Penthouse, making a successful adjustment from traditional pornographic media to "porno 2.0" – the era of digital media and social networking. With the Purchase of Various, Penthouse became the world's biggest adult entertainment company in the world. It projected revenue of $ 330 million for 2007, which was more than the annual revenue of its rival, Playboy Enterprises. However, compared to the other massive social networks MySpace and Facebook, which do not garner revenue through membership fees, this sale could have been even more spectacular. As Mr.Teamdating indicates in his comment about the Penthouse sale: "From a pure business play, porn is the only business on the Internet that works like normal business… How many members do you think MySpace and Facebook would have if they charged the same membership rates? Good guess…None."[5]

Before the sale took place, Various already employed about 300 employees and had launched twenty-five spin-off sites. After the sale to Penthouse, it became the world's largest adult entertainment network, owning a booming family of sex sites and a combined membership of more than forty million members. The network caters to multi-lingual communities based on various demographics such as age: *seniorfinder. com*, religion: *BigChurch.com, JewishFriendFinder. com*, ethnicity or nationality: *AsiaFriendFinder.com, IndianFriendFinder.com, Amigos.com, GermanFriend-Finder.com, FrenchFriendFinder.com, KoreanFriend-Finder.com, Filipino FriendFinder.com*, or dating preferences: *Passion.com, Alt.com*. The membership rates range between $ 25 to $45 per month, depending on the type of membership (standard, silver, or golden) or the length of the service. Standard members are non-fee paying members who are allowed to create site profiles but have very limited privileges, They do not get access to the AFF applications, including photography albums, voice and video introductions, astrology matchings, games, featured listings, rating systems, chatrooms and blogging areas. All these services are presented to fee-paying members through daily messages and ad campaigns.

In Hong Kong the site is available in the Chinese and English language. It became very popular with Chinese and non-Chinese women and men around 2005, individuals who were eager to enter the sex conquest as porno 2.0, despite the prevailing sex-phobic climate amongst religious groups and education platforms in traditional Chinese culture. The site had 60,000 listings when I started my research in August 2006 and almost tripled its membership by the time I ended it in August 2008. Other Western cultures that have a population size similar to Hong Kong have attracted larger memberships, but the Hong Kong site has nonetheless spawned a substantial group of web users. The site is primarily a heterosexual sex site for men and has attracted only a small percentage of female members. In December 2007 there were about 100,000 male members in Hong Kong versus 8000 females. This gender ratio remains constant in most other cultures, as there still is a global shortage of women who wants to sign up for this kind of website. Any woman who opens a profile is automatically swamped with massive amounts of requests from males, while males may be starving for a reply for weeks on end.

AFF.com can be seen as the sex district of social networking, but its multi-lingual and multi-cultural membership is increasing so quickly that it has become a mainstream network rather than a special interest group. It differs dramatically from mainstream social networks such as Facebook and MySpace by catering towards sexually explicit images and discourses that would be immediately banned or removed from these sites. Of course people can always sneak in sexually explicit materials on almost any network, but officially sex images are prohibited on MySpace. For instance, when creating an account and uploading a naked picture on MySpace, web users get the following photo policy message: *"Photos may not contain nudity, violent or offensive material, or copyrighted images. If you violate these terms, your account will be deleted."* [6] One of the main reasons for maintaining a strict obscenity standard, is a fear of offending conservative groups and losing business. Friendfinder Inc. is the network that tries to make money by going the exact opposite direction, i.e. by building the perfect social-sexual networking district.

Lizzy Kinsey Profile and Search Methods

I tried to engage with people as embodied individuals and as web agents on the AFF.com site. These sex spaces are not totally imperceptible as they are accessible to visitors on the web, but people customarily hide or modulate their embodied identities. In August 2006, I created and uploaded a profile for one Lizzy Kinsey, a 40 year old Caucasian bi-sexual woman. Alongside my scholarly ambitions and anxieties to get the research done, I also wanted to experiment with my own pornographic self-presentations. I wanted to show some "real flesh" to attract the AFF.com netizens and I took a range of pictures of my naked body. The picture I selected shows a close-up of my naked torso and breasts, while I am sitting down on my knees with have a pen lying on top of my legs. I have scribbled a written text on my stomach that reads "Are you Ready?" I meant to cast Lizzy as an outgoing sexually active female who uses a pen as a reflective tool. The pen could be seen as an instrument to play with or to record stories, but it was mostly ignored by my suitors who directed their gaze towards my stomach and breasts. Lastly, I followed the AFF.com fashion of the day disguising my identity by cropping the picture and not showing my face.

Aff.com profile picture Lizzy Kinsey

In the written part of the profile, I asked people to send me their erotic secrets and stories, or to meet with me and write some words on my body. The response was overwhelming. I received about five to six invitations on a daily basis. People sent me messages that contained extensive written profiles, one or two photos, and an initial introductory message. Some also attached their sex stories. I changed Lizzy Kinsey's profile a couple of times but I always kept it generic and ambiguous so that I would not get too attached to it and people would not understand its motivations at first glance. I would interact with people who appeared to be friendly and interesting. I slowly revealed to them that my real identity was that of a researcher who wanted to meet them for an interview concerning their experiences using the AFF.com website.

Most people simply disappeared at that point, but some remained and were willing to share information and engage in a face to face meeting. Even though I told them that I did not want to have sex, most people ignored that part of the message and still assumed that we would end up in the bedroom after the interview. I would usually arrange a preliminary meeting, conducting most of the interviews in public spaces. People kept asking to meet with me in quiet hotel room so that they could not be overheard. Some were more honest and suggested that we could do an interview while have sex. I did consider this 50/50 deal, but I realized that I just could not motivate myself

to have sex for personal reasons. I do agree with my interlocutors that I might have assembled richer data and more fulfilling dialogues if a sex session had been part of the encounter and methodological framework of this project.

The online exchanges were all conducted in English. 45% of the respondents came from Chinese or ethnically mixed couples, while 55% came from Caucasian couples. I also kept a record of selected messages and stories and archived a sample of one hundred messages received between July 15 and November 15, 2008. When dissecting this sample, I found I had received about 75 invitations to go on a sex date, 5 of which were sent by white or racially mixed couples, 38 by Caucasian males and 32 by Asian males. Many of the invitations came from Asian and Caucasian business travelers who would be visiting Hong Kong and wanted to go on a casual date. Additionally, there were Hong Kong-based expatriates and Chinese locals who wanted to date me. Since I received about five sex invitations per day, I immediately realized that there was a surplus of males and a real shortage of women on and the site. Even though my profile also requested dates with females, I received very few responses from women, but I did manage to get to know Ning of the duo "Double Trouble".

Most of the suitors were males who wrote rushed message asking to meet with a quick compliment about the photograph. They sometimes included their cell phone number or hotel information, which is actually forbidden and may be removed by site administrators. AFF forbids personal information in the initial exchange. A small percentage of adult friend finders wrote more poetic messages or sent longer stories or engaged in longer chats and email conversations.

Most members sent me compliments about my body or my profile in order to request a date. Some members made great efforts to sell themselves by writing self-aggrandizing profiles. For instance, about 30% of males describe themselves as "very handsome" in their profiles, which is a description rarely used in female profiles. Males also happily self-objectify themselves by sending a picture of their cock rather than their face or full body. About 50% of males used an image of their erect cock in an attempt to seduce Lizzy Kinsey. Additionally people took on clichéd alter-ego names like "EndlessFun", "HK juicy lover," "Black Stud", or "HK_puppy_dog." People seemed inclined to use the full clichés of sexual potency in pornography and used the penis image as the prevailing signifier of hungry masculinity, even though it was represented in a variety of shapes and skin colors, and in the many states of arousal from flaccid to fully erect.

Peter Lehman has argued that user-generated images of the penis on adult sites have significantly impacted representations of the male body and the penis. The male body and the penis have been excluded in the history of pornography, controlled by porn industries and focused on the representation of female bodies and vaginas. Lehman applauds a trend towards autonomy and diversification, where everyday web users define male sexuality. As he writes: "… it gives people who normally have no voice in these profoundly important issues, no conferences to go to, no journals in which to publish, no university presses vying for their next book, no students tying to get into their next class, a place to participate in this process." [7] My sample of images would endorse Lehman's point, as 50% of profiles contain an image of the penis, but members do make an effort to show the genital in unusual angles and colors to make the viewer more aware of the fact that there is no such thing as an ideal penis shape. Nonetheless, the surplus of cock images in my mailbox also made me feel alienated from actual persons.

Most of the stories sent to Lizzy Kinsey were detailed descriptions of the sex act which focused on genital intercourse. Other stories were more confessional and indicated that people were sharing them to reflect on sexual experimentation. Many of my conversations were ephemeral and did not lead too much, but some people provided interesting details. One man got completely turned on by the suggestion that I undress him and use my pen to cover him with words, but disappeared when I suggested a place and time for this encounter. Another curious proposal came from a 35 year old Caucasian male, who wanted to have sex with a woman and a male-to-female transsexual. He related his desire to the image of the pen on my thighs, which to him meant that we could both write ourselves into the other gender. He wanted to experience the shifting boundaries of the gendered body and male identity. He wrote

> You know – I have always wanted this but have I been afraid to try. I question if it is really too extreme, but yet the desire and excitement behind it makes me want to overcome that fear and try it out. Giving you a blowjob? Not sure, but I would try. Having a woman with a penis fill me up and come - will that make me feel what it's like for a woman to be taken by a man? And how will that feel? hmmmmm...is fantasy ever best played out or is it best to keep it in the realm of fantasy only?" This was a very good question and I was tempted to take it a step further, but this man also disappeared before we could continue the experiment.

Another reaction came from a bi-sexual married Caucasian man who had just started to visit gay saunas in Hong Kong and wanted to confess his experiences. I received many emails from him as he wanted me to also break out of my marriage and have sex with a woman. As he suggested:

> Wouldn't it be nice to think that if, after a few drinks with friends, you could hop in a cab and, rather than head home to DVD's or the little buzzy friend in the bedside drawer, get dropped off in a side street, go through a reception area to a dimly lit private lounge full of other like-minded girls? Somewhere where you get undressed by other girls and they lead you by the hand to the central play room which is all cushions and you fall gently backwards onto the cushions while they begin to touch you and massage you and get you to give them access to every part

of yourself - then return the pleasure until you decide to either sleep or go have a shower with someone and head home? No names, no questions, just pleasure. You step out onto the street several hours later, hail a cab and no one knows or cares... Now there's a concept.

I tried to correspond with this man and to convince him to share experiences, but he was he too scared to further reveal himself and was more interested in a furtive sex act between us. I had lunch with him once and tried to convince him to stay in touch, but he had no interest in my research.

Cleaning and Sharing Raw Data

Some AFF.com members were adamant about not being included in my research experiments. I found the following message on one member's profile:

> Any institution using this site or any of its associated sites for study or projects - You do not have our permission to use any of our profile or pictures in any form or forum both current or future. If you have or do, it will be considered a serious violation of my privacy and will be subject to legal ramifications. It is recommended that other members post a similar notice as well.

Even though I respect this member's decision to issue a warning message, I also believe that sex and media researchers have to be able to do their work and become more adept in analyzing online behaviors in a consensual fashion.

In developing a theoretical perspective on these and other interactions, Ken Plummet's notion of ethnography as a "reflective and reflexive hearing of voices" is a useful tool. [8] Rather than working with methods of structuring and filtering data, researcher can engage in mutual story-telling as acts of participatory ethnography. In *Telling Sexual Stories: Power, Change and Social Worlds*, Plummet argues that the circulation

and sharing of stories can structure the research and reveal social-sexual tenets and anxieties.

In the anthology *Online Matchmaking*, several writers have given accounts of the psychological ramifications of Internet sex and dating. McKenna connects many of these cyber sex experiences to the lineage of online role-playing games, where web users connect with each other for reasons beyond sex and romance, spending a very long time online as remote agents who slowly develop a relationship by sharing an online passion; a passion that may or may not lead to sex in real life. [9]

If we place AFF.com in this lineage of Internet sex, we can see that its apparent mission is to get people laid in the shortest amount of time, is still a novel sexuality. But AFF.com also comes out of long-standing history of personal ads in American papers and magazines. With the adoption of Internet platforms and a smooth web 2.0.interface, however, the site no longer provides news items or magazine content like its "hook-up" predecessors but expands the personal ad into user-generated content.

Masculine self-representations on AFF.com are focused on a horny sexual body and its hard genitals. These profiles can be seen as cheeky instances of self-objectification, casting the male as sex object. But they also simply echo the blunt macho rhetoric of the AFF ad campaigns. Daily messages target a member's frustrated libidinal energy; the entity who is stuck at home, rather than happily fornicating. The site is overall a source of frustration for women, who are turned off by clichéd self-representations and the lack of connection. From my own observation as Lizzy Kinsey, I found that most members are very direct or impatient and rushed in their interactions and would almost want to skip their time online.

But the case-study has also generated plenty of positive testimonies concerning sex affairs and sexual experimentation. The case study shows that people in Hong Kong are able to use the site to reinvent with old Taoist rituals as hyper-modern dating scenarios and life-styles. Even though the AFF site is owned by an American corporate network that has no interest in Chinese doctrines, both females and males have used the site to escape from domesticated life-styles. The AFF ad campaigns construct sex seekers as euphoric and competent agents who buy memberships and smoothly seduce others. This adagio is taken up by individuals and couples who use the site to brush up their sex lives or to challenge traditional sex roles. At this point in time, it is overall easier for Caucasian males and Chinese females to carry out this kind of pursuit, which is enmeshed in Hong Kong's history of colonization.

But the dialogues and stories show that the site also does allow members to break or reinterpret the scripts of historical sex culture, and use the suggested image of a smooth and competent operator to arrange invigorating affairs.

References

1   D21 press release. 2008. D21 Gallery, URL(consulted Jan. 2009): http://www.d21-leipzig.de/porno+zwei+null+english

2   Benjamin, Walter. 2001. 'The Work of Art in the Age of Mechanical Reproduction. In Meenakshi Gigi Durham. (ed.) Media and Cultural Studies: Key Works. MA: Blackwell Publishing Ltd.: 48-70.

3   Debord, Guy. 1995. The Society of the Spectacle. New York: Zone Books.

4   Ludovico, Alessandro and Hans Bernhard, (2007) 'Google Will Eat Itself,' In *GWEI*, URL (consulted Feb 2009): http://gwei.org/index.php

5   Hopkins, Jim. 2007. 'Penthouse' makes $500M hookup with social site Various,'In USA Today, URL (consulted Dec. 2007): http://www.usatoday.com/tech/techinvestor/corporatenews/2007-12-12-penthouse_N.htm

6   *MySpace Terms of Service* (2008), URL, (consulted Jan. 2009): http://www.Myspace.com/index.cfm?fuseaction=misc.terms

7   Lehman, Peter, 'You and Voyeurweb: Illustrating the Shifting Prepresentation of the Penis on Internet with User-Generated Content,' In *Cinema Journal* 46:(4), summer 2007: 111.

8   Plummet, Ken. 1995. *Telling Sexual Stories: Power, Change, and Social Worlds.*(New York and London: Routledge, Preface, XI

9   Mc Kenna, Katelyn. 2007.'A Progressive Affair: Online Dating to Real World Mating,' In Monica T. Witty, Baker and Jams A, Inman (eds.), *Online matchmaking* New York City Palgrave: 121.

Bonni Rambatan

OF INTERCOURSE, INTRACOURSE, AND THE ONE

Let me begin by quoting Grenzfurthner's review of Iain Banks' fictitious sexually-enhanced civilization:

> Scottish SF author Iain Banks created a fictitious group-civilization called "Culture" in his eponymous narrative. The vast majority of humanoid people in the "Culture" are born with greatly altered glands housed within their central nervous systems, which secrete – on command – mood- and sensory-appreciation-altering compounds into the person's bloodstream. Additionally many inhabitants have subtly altered reproductive organs – and control over the associated nerves – to enhance sexual pleasure. Ovulation is at will in the female, and a fetus up to a certain stage may be re-absorbed, aborted, or held at a static point in its development; again, as willed. Also, a viral change from one sex into the other is possible. And there is a convention that each person should give birth to one child in their lives. It may sound strange, but Banks states that a society in which it is so easy to change sex will rapidly find out if it is treating one gender better than the other. Pressure for change within society would presumably build up until some form of sexual equality and hence numerical parity will be established. [1]

A careful reader cannot but notice the strong Foucauldian overtones in Banks' imagined ideal society: the ethics of self-culturing, the reappropriation of the use of pleasure for one's own instead of it being used as an apparatus of power in a society of control. The benefit of such sexual enhancement – and perhaps one finds this unusual – is in the end, as stated, to "rapidly find out if it is treating one gender better than the other." In Lacanian terms, this possibility is its surplus-enjoyment (the actual/practical enjoyment – use-enjoyment, if you will, to further echo Marx, which I believe he rightly deserves – being the control of the body for sexual pleasure itself), and the final form of an ultimate gender equality, the substance from which such a possibility is derived, is its object-cause of desire.

Lacan is thus further confirmed when he claims that desire is always a desire of the other, and here we can even take it to the next level: our perverse, selfish desire is nothing but an introjection of a noble cause, which is precisely that of gender equality.

Kant with Sade, Posthumanized

The fantasy underlying such a statement is not a cheap and dirty one as one may suspect on the outset. It is not that "Oh, let us fuck in various different ways as we like, because in the end we are all good, noble human beings who strive for gender and sexual equality, anyway," but precisely the inverse is true: perversion is the price to pay for gender/sexual equality.

This is a claim that needs justification right away. How is it that having a good thing (gender/sexual equality) must be paid with having another good thing (wild, technologically enhanced sex)? Isn't there something wrong with this logic, something too good to be true? Aren't we supposed to pay stuff by losing other stuff?

Truth be told, the logic is indeed as such, and there is nothing wrong with the picture. It is rather the way we perceive the things as "good" or "bad" that matters. We must be very careful here: is it really true that such a wild and perverted sex, enabled by enhanced bodies and superhuman control abilities, is a "good" thing? Already, there are few who would respond with "Fantastic!" instead of "Bizarre!" This means that perversion, when taken to the next level, in fact leaps outside the Pleasure Principle, into its "beyond".

We can here return to Lacan's reading of *Kant with Sade*. [2] The point to be made is that Sade is not just another perverse pornographer, but an *ethical*

---

1 Text can be found on http://www.monochrom.at/arse-elektronika/about.html (accessed on February 12, 2010)

2 Jacques Lacan (1966): Écrits. New York: Norton. 2002.

philosopher. The rape acts of Sadean subjects are nothing selfish or even pleasurable – it is not something they merely desire to do, but something they *must* do for the great cause of Absolute Destruction. Sade's seeming perversion is in fact his radical ethical injunction.

Thus, is not Banks' imagined superpower of Culture – its ultimate freakish bizarreness – nothing but a price they must pay for the great cause of sexual equality? Culture can thus be read as more than just another bizarre exploration of human perversion, but an attempt at thinking an imaginary society in which things are better off, according to an Ideal, and what difference that such a society demands, the answers to which are: body modification, sexual enhancement, and superhuman control of our bodies and emotions.

If it can be found out that such sexual superpowers are what sexual equality demands, then it can be deduced that sex as we have it now (or more precisely the lack of control in sex) is precisely that which creates the gap in the Symbolic (the current functioning of our universe). Ideal societies always imagine a functioning, all-encompassing Language, a register that is able to manage, make sense of, and foresee all aspects of our life, a big Other, be it science, God, or otherwise. Thus it is only natural if we perceive that the ideal of sexual equality must involve the closure of this gap.

The current situation is one in which sexual equality is something yet to be achieved. The fantasy underlying Culture – and I believe underlying every other explorations of perversion and enjoyment, including those from Sade up to Foucault – is that the cause for this is the lack of control we have in our sexual lives. Sex presents us with a Symbolic gap in which a part (or parts) of us that we can never fully control nor understand gives us pleasure, alien and intimate at the same time. This needs no explanation as it has been the core of most dick jokes since the dawn of humanity.

Perversion, on the other hand, is the denial that the mother's phallus is missing. This coincidence is natural: it can be said that the Symbolic is incomplete inasmuch as the feminine phallus is nonexistent.

In Sade, the feminine phallus (or, more precisely, those possessing it: the sublime, "complete" bodies) can be found in the Sadean subjects and victims, who, after tons and tons of rape, are still able to survive intact. In Banks' and today's posthuman perversion, the sublime bodies take the form of bodies that shift and evolve, who, regardless of their sex and sexual activities, turn out to possess equal, ultimate sexualities beyond their physical bodies.

Post-Structuralism and the One-Substance

Such an imaginary notion cannot but invoke the notion of a complete language, in which full control in information, thus objective hermeneutics, management, and forecasting, will be possible. In the previous section I have shown you how Banks' version of posthuman perversion exemplifies a desire to complete our hitherto "lacking" language of sexuality with technological enhancement of bodies and mood/sensory control.

This, of course, remains loyal to the general postmodernist notion of shifting bodies and relative languages. Postmodernism, through post-structuralism, views the world as one purely and perfectly structured by languages, in which every object is a linguistic construct, with neither truth nor objectivity beneath it. However, precisely as such, the celebration of diversity it conducts is therefore, as Badiou has noted[3], none other than a celebration of *different languages*. Ideals can thus also be constructed through the creation of *new languages*, a notion which echoes from the Nietzschean Übermensch up to the Foucauldian self-cultured subject and beyond. Constructing

3   Alain Badiou (1988): Being and Event. London: Continuum. 2005.

new truths becomes easy, precisely because there is none, so it entails no more than a language game.

The danger of this school of thought is that while it purports to celebrate diverse forms of truths, it in fact still conforms to One, that of a Substance which can be constructed and deconstructed and reconstructed in your common post-structuralist language games but would in the end remain intact. In other words, it deals with different *representations* (realities) and *representation techniques* (languages) while maintaining one form of *presentation* (Substance) and one form of *representation method* (voice/text/images; "power structures" are constructed in the same way throughout time).

Is not what we are attempting today precisely a construction of a new sexual language, enhanced with all the possibilities offered by technology? If we are attempting a gender and sexual equality, is it not too premature to assume such a notion to be an objective, constructible one and that all we need is a new language, in the form of new bodies and new abilities of control, to access it? Does not such a notion of an already-present ideal, waiting for us to access it with our newfound abilities, echo too much of our monolithic, restricting pasts? Do we even know already what we mean when we say we desire "gender and sexual equality"?

Again we must be very careful and not fall into the temptation of the One. For we can already easily see its ruins once we bring in the more extreme players into the game: equality for scat sadists? Or (currently perceived to be) worse yet, for pedophiles? Or perhaps for the other extreme: equality for conservatives? With no ground of past ideals to be faithful to, it becomes difficult to map out trajectories of resistance, because they keep crashing with one another. The construction of new boxes – names, properties, parameters, categories – usually follows soon after to resolve such crashes, but of course – as the drama of the inadequacy of the State since its roots in the Greeks already shows – no box is ever going to spell out an "ultimate equality", for there will always be things left outside.[4]

Truths

It seems as if we have reached an impasse, in which every perversion-qua-ethical attempt for sexual equality is bound to fail. Is that indeed the case, are we left with a desperate condition in which we should abandon every attempt at equality and "lower" our sexual motives into that of simple masturbation and innocent gangbangs with the latest toys? While that may be possible to a certain shallow level, I doubt that removing such an ethical object-cause of desire from the crucially perverse scene will leave intact its previous field of effects.

I must admit that I have not done justice to those explorers of the possibilities of human bodies, senses, moods, and sexualities. If I criticize them, it is because I highly admire their courageous imagination and dedicated labor, and I have faith that they possess yet more potential than what has hitherto been credited to them.

I talked about avoiding the temptation of the One. What I mean by this is simply that our explorations should not center on a monolithic concept of equality that we personally believe objective, but it does not mean that we need to cross-check and confirm this notion of objectivity by positivist accounts and methods of creating new names and taxonomies to confirm our beliefs and accommodate more opinions. Instead, as a good Lacanian, what I believe we should do is to abandon the concept of objectivity whatsoever and instead fight for new, subjective truths.

This is not as abstract as it may sound. When we had our sexual revolution in the 1960s, it was not

---

4   One may go as far to say that boxes are in fact always-already constructed based on "things that will always be left outside", of which the latter is not an accident but a prerequisite, a foundation-qua-void, but this is another story.

a *reflection* built on some kind of perceived objective truth. Rather, it was an *act* of protest against the State, and it took the form of the creation of a new *lifestyle* instead of a political *science*. It was more like the action of the Braque-Picasso tandem rather than an art gallery's bureaucracy: chaotic and seemingly groundless safe for its ties to Cézanne and primitive art, which they subjectively claim to be a new form of aesthetics for which, at the time, had no established objective assessment criteria whatsoever. It showed militancy, not expertise.

I believe that sexual and corporeal technological explorations such as the ones presented at monochrom's Arse Elektronika conferences have potential to be tools for such revolutions. What we need to be aware of is that sexual bodies are not mere languages: they are also inert presence; not mere representations but also presentations.

Explorations of sexual bodies should not deceive us to think of them as mere variations of one single sexual Substance. There is no Substance, because there is no such thing as *a* sexual Truth. Because there are not only bodies, genders, and sexualities, but also infinite multiplicities of sexual truths.

Let us explore deeper than ever before.

Kyle Machulis

BIOMETRICS IN THE BEDROOM

The Boring Bedroom

At the moment, when most people think of adult toys, the control and integration systems are ridiculously rudimentary. The user is presented with a knob, a switch, or some other sort of coarse controller, which they are then expected to operate with some level of accuracy in a highly distracting situation. While this works on a basic level, why not use the most informative signal of all for control, our own internal biometrics? No more having to figure out which way is faster or what speed to go, just let the technology do the work and enjoy the ride.

Biometrics

The word "biometrics" can be used to describe many things pertaining to the state of a person:

- Identification/Security - Using what we are to identify who we are. Retinal scans, fingerprints, etc.
- Medicine - Using what our body says to learn about how we're feeling. EEG, EKG, Blood Pressure, etc.
- Control - Using data taken from biometrics equipment to affect things around us, from video games to large machinery.

In terms of what a sexual biometrics user would be interested in, we'll be focusing on the control aspect of biometrics. Usually identification and security have been taken care of before a situation where bedroom biometrics would be used, and we just consider both participants to be in good health (though the "you're fucking too hard and may cause a heart attack" monitor could be a viable business plan in the future).

Control and Loop Closure

Control Loops document what happen in a system when a certain part is changed. For instance, when someone turns a simple light switch on, the light is on until the switch is turned off again. This is considered an "open" control loop; there is no change to the system based on the state of the light (other than burning out at some point, but that's being ignored here).

However, when someone turns an oven on, the temperature rises to the desired level, then the oven turns off until the temperature lowers to a certain point, then turns back on, so on and so forth. The fact that the machinery uses the current state to feed into the next event is called a "closed" control loop.

Using biometrics for sex toy control allows the user to have what is known as "control loop closure". This means that, instead of having a semi-open control loop, like a user does when turning a knob based on the reactions of their partner, the control is dictated completely by the toy and the data it receives from the user's body.

Metrics

There are many metrics available from a user's body that can be used for toy control. We list a few examples here, for implementation ideas.

Pulse
Heart Rate is a rather obvious indicator of excitement level. As the user becomes more excited, their heart rate increases. However, it's coarse enough that it's really only useful in combination with other metrics, as there's no context to the number other than "body is pumping blood quickly". The hardware has no clue whether this is from excitement, fear, nervousness, or other emotions.

Galvanic Skin Response
Galvanic Skin Response is the measure of electrical resistance between two points on the skin surface. Since liquid decreases skin resistance, the definition of this metric is really "how much the user is sweating". Due to the massive changes in resistance from even the slightest change in moisture on places like fingertips, this is a way to tell how nervous or relaxed the user may be. E-meters and polygraphs use

this metric extensively, though they are not possibly the most trustworthy sources of information.

Blood Oxygen Levels
Using a pulse oximeter, hardware can obtain information about the amount of oxygen saturation in the user's blood. Not only can this be used to measure pulse, it also gives the hardware information on how well the user is breathing via how much oxygen they're taking in. This data can then be used to extract data about relaxation levels.

Temperature
Bodily temperature, especially in concentrated areas such as the groin, can be a great way of monitoring arousal. Blood flow increases during arousal, which causes the temperature of a certain area to climb.

Brain Activity/Electroencephlogram
EEGs measure the electrical current on the surface of the brain. Through advances in our understanding of the brain, this information can be used to tell many things about the user, from meditation and attention levels, to whether or not they have certain diseases related to electrical issues with the brain (i.e. epilepsy, Crutchfield Jacob's syndrome, etc...). In recent years, video games have started using the ability of the user to control coarse signals in their brain by thinking certain thoughts in order to control video games.

Plethysmographs
Penile/vaginal plethysmographs detect the amount of arousal in the user. They consist of a condom like object that goes around the penis, or else by monitoring the inner skin color of the vagina (a photo-plethysmograph), measuring blood flow. Many psychology experiments have used plethysmographs while showing images to a viewer in order to track their level of excitement. This could be used for other situations, though the structure of the device does not make it very useful for anything involving direct stimulation.

Pelvis Pressure Monitors
Usually a vaginal or anal insert, pelvic pressure monitors give the user an idea of how much force their pelvic floor muscles are producing. This is usually a great indicator of orgasm timing and strength, as well as a way to exercise and tone muscles that can be used for sexual pleasure.

Hardware

To be able to obtain the aforementioned metrics, hardware must be worn by the user that can communicate with the computer. As of this writing, there are many different hardware solutions available on the consumer market. The following are just a few examples.

Polar Hardware
Polar, a sports hardware company, produces a heart rate monitor that can communicate wirelessly with a computer. And certain hobby electronic stores often sell breakout boards specifically for this sensor.

Lightstone
The Lightstone is a USB widget that records blood oxygen levels and Galvanic Skin Response. It was originally used with the Journey to Wild Divine game, the goal of which was to teach the player how to properly relax. The device can now be found for sale fairly cheap on the internet, and cross platform drivers exist that allow it to be used on any platform, and through many programming languages.

Video Game EEGs
In the past 3 years, consumer EEGs have been released for video game control. These quite often come with some sort of SDK that allows programmers to access the hardware, and can track meditation and attention levels, as well as providing a full raw wave from the sensors that developers can access themselves.

Usage Ideas

Simple speed control
One of the simplest uses of biometrics is in direct speed control of a vibrator or other toy. For instance, set the speed to rise and fall in time with pulse, meditation level, or similar signals, or average the change over time and use that to set speed. The programming is simple, but the effect is still quite intense, since it relates to something happening within the user's body.

Machine Learning and Timing
Machine Learning allows a computer, given a certain number of data sets about a situation, to extract information and patterns in those datasets. This means that a training program could be written to log the user's biometrics over the course of a sexual event, from warm up to orgasm. Assuming enough of these events are recorded, the biometric information could then be used to teach the computer how to be a "lover", or at least, a "teaser". This would allow the user to set certain time limits for a session of play with a computer, and the computer could then dynamically react based on the patterns obtained from the user's past sessions. This means that the program would have some view into how close the user was to orgasm, and could change outputs (speed, rotation, etc...) appropriate to move these levels in relation to the time limits given.

Situation Bondage
Similar to the speed control idea, the user's biometric information can just as easily be used against them. Pleasure could stop if pulse reached above a certain level or attention was not high enough. All sorts of interesting games could be created simply out of setting some limit over/under which the situation is changed.

Aesthetics

Most of the equipment mentioned so far is either for medical or video game use. In those cases, users don't so much care about the aesthetics as they do the end outcome. However, in a sexual situation, the look becomes possibly as, if not more, important than the function. Therefore, it is important to take into account the aesthetics of the hardware when integrated it onto, or into, the user. Sensors can either be hidden in lingerie, or, depending on the fetishes of the user, left open with lots of bare wiring.

Ethics

Another important factor when using biometrics in the bedroom is ethics. Once technology is integrated into the event, it's possible for users to rely on it too much, to the point of sometimes removing common sense from the situation. The ability to reduce the situation to numbers invariably adds a level of abstract between those involved, which must be heeded and adjusted for.

The Future

So far, biometric toys account for roughly zero percent of the mainstream commercial market in sex toys. Everything written so far in this paper is either happening in research or in the mind of writers, but certainly is nowhere near the norm for adult toy interaction methodologies. What we have outlined could be only the beginning of the next major wave in non-organic physical pleasure, or it could never see realization. It is simply hoped that someone will try, because that's the fun part, anyways.

Bibliography

Master John: Peripheral safety and feedback system
for practitioners of BDSM, Chris Noessel, Sexual
Interactions Workshop, SIGCHI 2006, http://www.
ics.uci.edu/~johannab/sexual.interactions.2006/
papers/ChrisNoessel-SexualInteractions2006.pdf

Introduction to Pulse Oximetery,
http://www.pulseox.info/

Biometrics, Wikipedia,
http://en.wikipedia.org/wiki/Biometrics

Control Theory, Wikipedia,
http://en.wikipedia.org/wiki/Control_theory

Plethysmography, Skeptics Dictionary,
http://www.skepdic.com/penilep.html

Saul Albert, Tatiana Bazzichelli, Violet Blue, Carol Queen, Eleanor Saitta, Douglas Spink, Rose White

OF HYPERCROTCH AND NANOBOT
ARSE ELEKTRONIKA 2009 CLOSING PANEL / Host: Johannes Grenzfurthner

Johannes Grenzfurthner: *(microphone squeaks)* OK. I think it needs lots of love I guess, OK. So, please sit down ladies and gentlemen. It is the last step though in today's conference. It is the grand finale. It is called Hypercrotch and Nanobot. Nobody on the panel knew about the title although it was on the website for like a month or something like that.
But whatever, OK? So...

Carol Queen: We saw the title. We totally all saw the title.

Johannes: Yeah, that's what you say now. My general idea about the panel is not to force you into a direction. I generally want to round up today's presentations and in general, I want to go back to the general topic of this year's Arse Elektronika: "Of Intercourse and Intracourse". So genetics, biotechnology, wet ware, whatever there is. The future of sex in the sleaziest way possible. Let me introduce our panelists. On the far right side and that of course is not a political thing, we have the wonderful Violet Blue. She is the sex blogger, sex celebrity and last year I misheard "sex educator" and understood "sex agitator". But I think "sex agitator" is actually way better. You should put that on your blurb on your site.

Violet Blue: *(doesn't use microphone)*

Johannes: ... and use the microphone. I am very hardcore anal BDSM about the microphone here.

Violet: Is it on? *(microphone squeaks)* Oh, very much so!

Johannes: Yeah, because we might actually get a little bit of funding of the Austrian government. But then we have to record all this stuff and put it online – as a proof! Next panelist is Eleanor Saitta. She was our keynote speaker today and she talked about the future of sex.

Eleanor Saitta: Yeah, the future of sex, the future of designing sex.

Johannes: We have Carol Queen. I mean who does not know Carol Queen? She is running the Center for Sex and Culture. She is a – to overuse the term – sex agitator as well. And she has problems with the Catholic Church at the moment...

Carol: Yeah.

Johannes: We can talk about that in later. The Catholic Church will be there anyways. Whatever future we can imagine, there is a Catholic Church...

Carol: Not every future.

Johannes: Blatant optimism.

Carol: I'm also the Luddite of the panel, people. Everybody is going to say really smart things and I'm going to say, 'I have to go look that up later.' [laughter]

Johannes: We have small gadgets. And we use them. For example to send a tweet to… Rose White!

Rose White: Hello.

Johannes: You are a sociologist.

Rose: I am a sociologist.

Johannes: And you are doing lots of research… you are a regular at Arse Elektronika. But last year, she actually wanted to give a talk at Arse Elektronika, but she had to undergo surgery. Bad stuff!

Rose: It was kind of an extreme body modification. I removed an organ but didn't do it on stage.

Johannes: And the organ almost was penis-shaped, an appendix!

Rose: It is an appendix. Instead of giving the talk at Arse, I gave it at the Chaos Communications Congress to a delightful standing room crowd in the main hall. So that turned out great.

Johannes: But we printed your paper in our Arse Elektronika anthology…

Rose: It is in a book, which you should buy.

Johannes: So thanks a lot Rose for being here. We have Tatiana Bazzichelli.

Johannes: You are based in Denmark, aren't you?

Tatiana Bazzichelli: I am Italian, but I am based in Denmark and now, I am living in San Francisco for four months because I am doing a Ph.D. research on Web 2.0 and hacker culture at Stanford University. So yeah, I am a nomadic entity.

Johannes: You are one of the heads of CUM2CUT porn film festival that was based in Berlin. Thanks for your participation. And we have Saul Albert and Saul Albert is a member of the faculty of "fuckology" of the Open University. Am I right?

Saul Albert: Close. The Faculty of Fuckology or alternatively, the Faculty of Physical Education at the University of Openess. The Open University is a very august educational institution in the United Kingdom. The University of Openess is its somewhat less reputable counterpart.

Johannes: Tell us a little bit about the University of Openess.

Saul: Sure. The University of Openess is a self-institution for research into methods of autodidactic education. Basically in 2002, it took some of the pedagogical principles of the open source movement and applied them to a liberal arts education model in an abandoned town hall building in East London. A whole lot of different groups converged around this space and started teaching each other all kinds of stuff. The Faculty of Physical Education started with a course on how to evade paying train fares by crouching and hiding in between luggage racks... I think that method was called the 'Extreme Fysical Fuk' and involved the use of masturbation to alleviate the symptoms of crouching in the cramped luggage storage area between two chairs.

So, anyway, it kind of evolved and turned into a relatively large scale and very diverse institution with I think about forty faculties and lots of different research projects, all kinds of stuff. Google University of Openess. You will find all kinds of interesting things. That was back in 2003/4. These days when 'social software' has mainstreamed as 'Web 2.0', and a lot of the aspirations from the early days of open source have been appropriated, I think many of the UO's assumptions have been proved wrong, which makes it more interesting to me. Anyway, the UO has been in remission for a couple of years, although it may be about to turn malignant and start doing stuff again.

Carol: What is the difference between the University of Openess and a sex cult?

Saul: I don't know. I have actually never been part of the sex cult so... we'll have to swap notes.

Carol: That is a sign of indoctrination actually, when they say they don't know.

Johannes: OK, Tatiana.

Tatiana: Let me draw parallels between open technology and pornography to connect to what you said. And also, to explain deeper what you mentioned about the CUM2CUT festival and to give some examples as input for future discussion. We defined this festival as a kind of open pornography event. It actually was an occasion to develop a network of people who wanted to freely express their sexuality far from mainstream porn. Open porn because it was a platform to share sex strategies, art experimentation, irony and pleasure and to transform technology into a porn tool. C2C, as a hack of P2P, came out to link pleasure and orgasm to the action of sharing pornography. I invented the festival together with Gaia Novati, who is also Italian and is living in Berlin. And somehow, for us, it was a kind of "open sexual university" in the sense that in three days people could learn from each other sharing sexual experiences. People had the possibility to create a short porn movie, in just three days, and also edit it at the same time, so we called it a Porn Marathon. It was addressing normal people who had never acted in the porn field, that answered to the call to make pornography in the way they wanted, but at the same time, to try to reinvent the concept of porn itself. So, it's open pornography because it's like an open concept that everybody can rewrite and personalize through different bodies, minds, and crossed-networking-experiments.

Johannes: Rose, I'm talking about the Western world or the Westernized world. So I'm not talking about Iran, where you actually get killed because you're a homosexual and awful stuff like that. So we have to always divide that in this like, open, open-ish, liberal-ish thing that's going on in the Western world. So, there's lots of talking about the mainstreaming of porn, there are many, many artists out there who use porn as their medium in a certain way. We talk about the future of sex, and we talk about the future of humanity. How do you see that field as a scholar in sociology, how does that fit together?

Rose: Speaking as an academic and speaking personally, what I see happening with consumption of porn and in younger cohorts, are that... picking up on the theme of Openess, that there is much more Openess and flexibility about, say, gender roles in porn. Or in sexual behavior and thus more Openess in looking at porn where there's fluidity. And so when I look at a young pornographer like Furry Girl, who's here, and some of her porn sites and the porn that she makes, I think about the people consuming that porn and that from a site like NoFauxxx where there is a lot of genderqueerness, and I'd like to think that the sex acts that are going on there and the people who are excited by them are part of the future that Eleanor is talking about, and that Annalee was sharing with us, the idea that we're sort of evolving away from a kind of dichotomous setup into just exploring what we enjoy.

Johannes: My question for Carol would be, isn't like 99 percent of porn out there boring? Heteronormative mediocrity? Is there a revolution on the horizon of porn?

Carol: Well, I'll tell you, there are two ways to answer that question, and the first way is, "Yes." [laughter] The other way is a little more nuanced. The other way to answer the question is: yes, but a lot of what's in porn is intensely erotically appealing to many people, sometimes only snippets, sometimes only certain elements, sometimes only certain performers. And as I think Susie Bright wrote about 20 years ago, women have historically looked at male-oriented porn and just sort of spaced out on all the other dreck that's part of the porn movie, and pulled out to focus on and watch what they *wanted* to see. I think that still continues to be true, although there are many more things now that are at least putatively women-focused, and it's desired that women will buy into them. Some women do, some women still don't find porn that they resonate with, as some men and transgendered people also do not find it.

But I think it's really important to say that some people always have, as long as there's been any kind of technology they could control, made porn *out of* mass produced porn, that was for *them*.

The Institute for Advanced Study of Human Sexuality, where I got my Ph.D., the Center for Sex and Culture, which I co-founded and curate, both have items in our collections where people used their record switch judiciously to pull pieces out of television commercials that they thought were hot - or they picked out scenes that they liked in pornos and looped them to make them longer - or things that the computer helped them create...

I said I was a Luddite, but I *do* know that computers exist, and I do know that the computer makes this process of auto-editing easier and will make it easier and easier over time. In fact, I wouldn't be at all surprised if there were some mass-market editing programs that really help people do that.

I know there are things that savvy people know about already, but I think, at the same time that kids are going to be able to make their lightsabers raise into the air and do backflips, there will be tools that their slightly older brothers and sisters and dads and moms can use to make the kind of porn that they want.

I think individuals desire erotic entertainment and they want to see reflections of themselves and their erotic interests in the media that they consume, so that's not going to go away.

The future, I think, is going to hold more people for whom that's true, especially because they have been able to go on the Internet and see alt-porn of various kinds - however you would want to define that, there are many ways of thinking about it. Because they know it's out there, there will be more markets that are growing to meet that knowledge. Good Vibrations, where I have of course been involved since the dinosaurs ruled the earth, is exploring some ideas around that, so you'll see those projects soon, and I'm sure we're not going to be the last... I don't even know if we'll be the biggest.

So, we'll see what happens. But I'm completely sure that as we have more individually easily usable technology, more media is going to be sexualized in a customized way. I just don't have any doubt of that at all.

Johannes: It was really interesting because as part of the film festival that was co-hosted by Arse Elektronika and monochrom and Blowfish, we showed a documentary from 2001 called "Love Machine" by Peter Asaro. And you were part of it. It was kind of interesting to see... it was nine years ago! We wanted to show the documentary because a) on a certain level it is completely outdated. Nine years can be a very long time in robotics. But b) many of the things that were said in the documentary were still pretty much like what we would actually discuss right now, a decade of years later. I remember that Eleanor was sitting in the back of the cinema and writing her talk for today, and every now and then I heard her outbursts of "Oh no! No!" [laughter]

Eleanor: Well, as far as the future of sex with robots goes? One of the things I found really interesting about that documentary was that specifically they chose to focus exclusively on people trying to build robots that looked like people and that worked like people. And I was really curious, where are all the people who want to sleep with robots that look like robots? You know? I mean...

Audience member: Internet!

Eleanor: Yeah, but it... [laughter]

Johannes: Carol wants to sleep with the Internet, yes, I know that. [laughter]

Eleanor: So I think that's one thing which will start to become interesting. So, once we get sex toys which are kind of entering into that uncanny valley territory, at what point do we stop trying to purely emulate? At what point do we say, "OK, I'm going to create something else that's still hot but which isn't trying to be quite exactly human?" One of the other things that we were talking a little bit about during the film was, you know, what does it look like when you build robots that are trying to have sex with other robots? Can you build... There's an art piece. Something else. Robots that are exchanging means of reproduction, that are exchanging their own robotic DNA with each other. Either as art, as something somewhere between a sex toy and art, or as something actually useful. Can you enable that kind of sexual style reproduction in robots? Is that a useful set of techniques to bring to bear for actual robotics work?

Johannes: It's almost like a Darwinian art piece. I don't know. I remember, there's one guy who founded Real Dolls. He's explaining that he started it as pieces of art, that he wanted to sell it on the art market, and nobody was interested in realistic looking, creepy, Real Dolls to be placed in the corner of the living room as human-like sculptures. But more and more people were interested in having sex with the Real Dolls. The guy portrayed himself as the suffering artist: "Nobody understands me. I want to make pieces of art and they just want to fuck it." [laughter] We were talking about the possibility of the future of art installations you can actually have sexual intercourse with, who do not need to look like female or male, android, whatever things. Just like, "I really would like to go to the Museum of Modern Art in San Francisco and fuck an installation there." [laughter] I kind of can see that, like an abstract kind of thing with a couple of holes.

Saul: They've actually got this piece on display right now, which is this wall that was made out of plaster and kind of bulging fabric that I actually ended up… It has lots of holes in it. It's, unfortunately, quite stiff, but you could make something like that that was a bit more flexible. Spray some lube on it.

Johannes: The new world, new fabrics? What do you think, Violet?

Violet: Thinking about all of that, thinking that we could actually make the better fuck machine if we're not trying to make it human is a really, really tantalizing concept. Not just because of all the creepiness of the uncanny valley, which is appealing for a subset of people, myself included. I was doing a podcast last night, and someone was asking a question about flavored lubes for a cooking podcast that we were doing. I was talking about when working at Good Vibrations and testing and tasting all of the disgusting flavored lubes that we carried…

Saul: I think I need a Bud Light now.

Violet: Yeah, exactly. So, for instance, this, be it lemon flavored… The dick is not going to taste like a lemon. It's going to taste like a rubber piece of lemon. It's going to be gross and warm. But we found that things that tasted like something artificial, like the flavor of bubble gum which was made to be an artificial flavor tasted fantastic, because it was made to be fake. So, that's sort of the concept I'm thinking of when thinking of what if we start designing fucking machines not to replicate, necessarily, the human to human experience but the human to robot experience, and how much more effective in orgasm and experience those machines could be.

And it also reminds me of last year. We were talking with Thomas Roche, talking about the appeal of fucking machines and why Fucking Machines as a site is so appealing and why there are so many female members, so many subscribers that are female. It's because of the pure relationship of the woman to the machine itself. It's not a woman and a man. It's not a woman and another woman. It's a woman and a machine, and she's getting off purely for her own pleasure. Something about that is exactly what works for a lot of people.

And also, it's HeartCore. HeartCore Productions is your new Good Vibrations line of videos.

Carol: HeartCore, Real Queer, and Pleasure-Ed.

Johannes: Microphone. Microphone. We need it for recording, goddamn! It's pure taxpayers' money from Austria! Every second!

Carol: The video names that Violet reminds me to say are HeartCore, Pleasure-Ed, which is my part of the project, and Reel Queer. And you wanted to say something, didn't you? I did too, but you were jumping first.

Rose: I was jumping first. So there was another guy in the Love Machine documentary who, while he was talking, had a mechanical hand that he was wearing.

Johannes: Like claws, almost claws.

Rose: Almost a sort of removable hand, and he kept making these kind of snikt-snikt Wolverine noises while he was talking. He had all this machismo while he talked as opposed to the RealDoll creator who had, as Johannes said, a sense of [sad voice] "You want to fuck my art." Whereas, the Wolverine fellow was like, [macho voice] "Yeah, I've made a robot for you to fuck!" [laughter]

Johannes: "I want to give pleasure to as many women as possible!"

Rose: "You know, I have been talking to my friends, and I talk to them about how much you would pay over your lifetime to fuck?" And so, he was asking them, in the times when over the course of things you have to pay... He was trying to figure out what was a reasonable amount to pay for a robot that you would fuck, and the robot that he had made was much less lifelike looking. It was... Although he gave it some humanoid female characteristics, he wasn't trying to make it as flesh-like as possible. It was much more science fictiony and robotic. I thought that it was interesting, although there were cultural differences between him and the other guy, and certainly personal differences, that he was much more enthusiastic about the sex, and it really was sort of resonating.

Johannes: Saul?

Saul: I was just going to say I think there's a flip side to this. I think it's fascinating, the idea of having sex with inanimate objects or animate objects that that aren't identifiably juicy, or plump, or nice to touch, or tangibly similar to human to human sexuality. In fact I think it's important that proper scientists and roboticians - people who are not just building cyberpunk sex gizmos, but biotechnologists, particularly, are educated in how to follow their libidinal impulses to innovation. I think that it's as important for them to be turned on by their creations and to feel that it's OK to follow that impulse as it is for people with a penchant for fucking robots to scrape up the litter of scientific innovation that they can access via the Internet.

Carol: Well, there's an argument to be made that that's the difference between a good and a bad sex toy. There are plenty of sex toys that are designed by people who don't have any idea about the human body, certainly bodies of other genders than their own. They are designing the sex toy because somebody is paying them to do it, but they don't use the toys themselves. Is it hot for them? A lot of them it's not, or they won't admit it. You don't get that evocation of hotness from the object that they create. So I think there are a lot of levels to this discussion, some of which are a little on the old-fashioned side, and obviously ones that have to do with technologies that

we have barely begun to realize can be deployed this way.

But, I wanted to say something else about the robot and the object thing, because I just yesterday got turned on to a woman who is a sexological researcher, and she's writing about something she's calling object sexuality. Her name is Amy Marsh, and I'm going to invite her to come to the Center for Sex and Culture one of these days. So, if any of you think this is a hot idea, please get on our email list so I can let you know about it when it happens.

She's looking at a cohort of people who mostly don't have desire for humans. They might, under other circumstances, identify as asexual, but they don't because they have objects they desire instead. Not all are objects that you would think of initially. Immediately, I thought of, "Ooh, Hitachi Magic Wands and dildos. Those are very regular in my world." No, she's talking about somebody who's actually married the Eiffel Tower, people who are erotic around objects that many of us would not necessarily recognize in our own minds as being erotic objects. That's pretty fascinating, and as soon as she starts talking about that in any kind of a wide-ranging way, more of those folks are going to show up and talk about it, and we're going to understand more ramifications of sexuality than we did before, because currently, people don't talk about this all that much.

And they resist being called fetishists too, that's the other thing that I thought was quite interesting, that that is the tendency of the "old school" and they don't relate to it. And sometimes maybe those of us who are comfortable with the notion of being fetishists, or being in a fetish community, say, "Oh, well, you've got a fetish then," and they're like, "No. It's my sexuality, it's not my fetish. It's who I am."

So, those differences are going to be really important as we march forward and have more excellent gizmos that can do more cool things for us. Or with us.

Johannes: Tatiana!

Tatiana: Yeah, I wanted to say something. I think that maybe a good strategy for the future would be start thinking about the present of sexuality, and the question is: How many people outside this room know and happily practice the topics we are speaking about, and what is their perception of sexuality? What is sexuality and pornography for them? I think that it's really fascinating to speak about sex with a machine, but first of all I wonder how many people are really confident with the topic of porn itself, which is their mental representation of pornography and if they consider porn as part of their everyday life. I found really interesting what Saul said. We need to stimulate people to think about sexuality and pornography as an open concept, as something that could belong to everybody, not just people who are into a specific community.

Speaking about the porn market, I think that actually thanks to technology and the Internet we lately arrived to a point where a lot of people are really sharing pornography, and they are transforming the porn market as well. This is because they practically act outside the market, in the sense that they use peer-to-peer channels, and systems to share their files, images, pictures, videos, etc. I would say that this is creating a parallel market because they are not basically generating any kind of capitalist revenue. What they do is mostly grassroots oriented.

Somehow we are able to influence the market, the one made of stereotypes that we don't like, - the mainstream pornography -, but we can do it starting to think about something different, creating a contamination of ideas and culture, promoting alternatives which concretely reach people. But this can happen mostly if we are able to bring porn out from this room and not just among people that are well-educated. It has to become part of our daily life to make a difference...

Rose: I really love the things that you're saying, Tatiana. I think there's no mistake that... it's not an accident that monochrom is involved both with this conference but also heavily involved with the hacker space movement, and with open source software, and open culture in general. So, as you're talking about works that are being produced outside of the marketplace, outside of capitalism, that is exactly like the things that are going on in hacker spaces. Which

are not generally pornography, although, tomorrow's workshops are all going to be happening at Noisebridge, which is San Francisco's main hacker space. [applause]

And one definition of hacking that a friend of mine gave is that it is the work that we do that is just for ourselves, that's not for money and it's completely outside of the capitalist system. So...

Tatiana: I really like that you mention hacker culture, because I come from Italy, and for us hacking is a "life attitude". I speak about at least the community I'm part of, that is the one that brings together art, hacking, and often sexuality. In the Italian background sex culture is really connected with hacking because it shows a challenge of trying to rethink sexuality in a different way that is not predictable, and it is not mainstream. It plays with boundaries and stereotypes. It is like a flow that is traveling between different identities, different ideas of using your body.

Metaphorically, you open your body like a hacker could open a machine, in the sense that you try to experiment with what is inside, with your deeper emotions, and you create a new mental category, or get rid of categories.

Johannes: Madame Blue!

Violet: Yeah, I just wanted to add quickly too, I think what you're talking about is not necessarily subverting but changing the idea of what porn's normative culture is, through allowing people to sort of open-source their own pornography, or open-source their own images into pornography. And as someone who's been blogging about porn since about 2001, the revolution is already underway. And a lot of that has to do with some of the things we were talking about earlier, and that's that the bar to entry has changed completely in terms of content distribution and content creation in porn.

And the DVD is dead and dying and thank fucking god for that. And what's great about that is that over the past few years, as this method of content delivery has died, slowly, people have started making their own porn and distributing it on the Internet.

And that's why we have all of these great, you know, women of color, queer-owned porn companies making their own porn, and yeah, in the background they have hot queer hacker chicks that work there that know how to put the sites together and know how to work with the video files and put stuff on the Internet. And this revolution has been going on, somewhat quietly, in the background while mainstream porn has just been churning out more and more of the same old stuff. And we're only now really starting to see it now that people aren't buying DVDs anymore and they're going online for porn.

So, the revolution in porn is open sourced, it is coming from the people, and you can find it. It's loud. It's out there. And it's really, really exciting. And then, the only other thing I wanted to add too was to Carol in terms of the woman talking about people sexualizing and having sexual and emotional and romantic relationships with objects.

It's not a new thing. I just recently discovered an arena of porn - it's Japanese - it's table-porn. It is exclusively about having sex with tables. And there's a lot on bit torrent, it's excellent. Most of it is clothes-on, and it's very, very long segments of girls having orgasms on the corners of tables. So, it's not a new thing.

Audience Member: Trip to IKEA tomorrow!

Johannes: IKEA! Yes! We go to IKEA and later on we go to Noisebridge, OK? So, we have one comment from the audience, Douglas Spink who gave an interesting talk today about cross-species science fiction literature, and... but he was first, he was first.

Douglas Spink: I have to take off for the airport in a minute, so I'm going to jump into the discussion before the official comment period begins. In terms of object sexuality, we've got a quite interesting thread at our company forum - www.cultureghost. org - where we started tracking this subject a while ago. It includes some really detailed investigations of a woman who 'married' a roller-coaster, which was reported recently. This story actually crossed over into them mainstream, into the conventional press - i.e the newspapers - and it was treated with respect, not labeled dismissively as a "fetish" as is so common with unconventional sexualities. It's approached in these articles as a sexual orientation, in terms of how we normally think of the concept of sexual orientation in a broader context - with "we" being those of us who are ourselves part of non-mainstream sexual orientations. In the same way we no longer consider gay folks to be "suffering from a fetish," we're slowly but surely seeing an increased recognition of other orientations beyond straight or gay. Orientations are orientations; they have the feeling of orientations - not fetishes - which of course does depend somewhat on where we're at when we judge what is and isn't a "real" sexual orientation. It's clear that orientations are much broader in category than a simple, one-dimensional gay/straight binary pair.

Further, in regards to the discussion about technologically-based items of a sexual nature - and, in particular, discussing items that aren't merely attempting to present themselves as replications of essentially human characteristics or components - no... it's not your imagination: this kind of thing is increasing common, and skittering towards the mainstream. Example: Penthouse magazine just this month ran an article about a company, based in Scotland, which makes high-end sex toys. These toys aren't motorized - yet - but they're actually working on that sort of thing currently. In any case, what's interesting about these toys is that, whilst they're still essentially gendered (sort of...), they have absolutely nothing to do with being simulacra of human *anything* - in fact, they're not even loosely based on anything even vaguely mammalian. They're completely, 100% fantasy sex toys - in the most fundamental sense of the word "fantasy," i.e. something that doesn't exist and never has existed in the physical universe. They're very high-end, not just cheap prop pieces or whatnot. I actually did a bit of informal consulting work with the folks who've created this little company - a few years back, we ended up being in touch through various colleagues known in common amongst our community. Back then, it was really more or less an idea, a concept... since then, it's really exploded. I mean... it's really exploded: they can't really keep up with demand in terms of their ability to manufacture products fast enough. And now Penthouse has done a full-blown article on them - a front-page article.

Violet: Can you say what they're called?

Douglas: Oh, the company's name is Bad Dragon; website: www.bad-dragon.com.

Violet: Oh, yeah.

Douglas: You've heard of Bad Dragon already?

Violet: I've blogged about them. [laughter]

Douglas: When I was first talking with the founder of Bad Dragon about the concept, we were more or less joking around... "Wouldn't it be cool to do these pure fantasy, high-end sex toys?" And it's just gone from there!

Violet: They're silicone toys that are in the shape of fantasy creatures, like, dragon penises that ejaculate.

Douglas: Some of the shapes of the toys aren't even pretending to be creatures at all - fantasy or otherwise. It's as if they realize at this point that there's no a priori _need_ to have a creature attached to the concept of the toy itself. They are really shapes - shapes that are appealing, that customers desire from an aesthetic standpoint. I was, personally, seeing these things as more or less decorative stuff, initially - I'll admit it; people would simply buy them and not really 'do' anything with them, just put them on the shelf to appreciate visually. Indeed, some of these things are honestly huge, ok. So I thought folks would be putting these $300-400 objects d'art on the shelf and just looking at them. Nope... oh no: they *use* them, they use them so much that they *break* them - and they want new ones to replace the broken ones! And the interesting thing is that they sell a lot of them... like, a lot of them. They just can't keep up with demand, from time to time. They don't really do any serious advertising, in terms of spending money - they've never been dependent on advertising. And, in point of fact, it's not just the furry community as one

might expect; there's a corner of it that's the conventional furry community, but it's much more than that. It is a post-humanist sexuality, an example of that - and that concept is one that, as I usually say, I can "smell coming from a mile away." It's so close, it's basically here already. It's in Penthouse magazine, for Dog's sake - they just wrote it up! I really couldn't believe it, when I saw that kind of coverage for something that'd be more or less unimaginable even 10 years ago, in terms of a real company with real products and real revenue. Unfortunately, I can't really dig into any of the deeper issues surrounding post-humanist sexuality, given the time constraints, but I did want to say that from where I'm sitting you guys are ahead of the curve - or at least right on the cusp of it - in terms of your awareness that there's something going on that's much broader and more interesting than past efforts to mimic human sexual elements with technological elements. That's already been done, and isn't where the powerband is in the future.

Indeed, the curve has caught up with where you were even a few years ago, and that curve is carrying forward at an accelerating - perhaps geometric - pace. Indeed, even I find it hard to keep up with the leading edge of that curve now - and I'm quite well-qualified to be aware of such things! I keep my ears open for interesting things and I hear about stuff and I think "ok, wow, that's weird and interesting and unexpected - even for me!" And then I'll get an IM or tweet from someone at Bad Dragon and, sure enough, they'll have something entirely new in their product pipeline and I'll think "wow - who came up with that?" And it'll go into production and it'll sell - and the process keeps accelerating...

In terms of non-mainstream distribution channels for explicit content, in a similar vein you guys are absolutely right in your observations during this AE conference. As an example of our perspective, our network security company - Baneki Privacy Computing (www.baneki.com) has been making use of a bit torrent-based 'tracker' as an internal testing facility, in terms of ensuring our privacy services work well with BT-enabled file sharing protocols. We just needed an internal tracker backend, so we can make sure it works well with the services we provide, specifically TorrentFreedom (www.torrent-freedom.net). And at some point, outside folks started throwing stuff into the tracker database index, some of it vaguely cross-species in nature but mostly an eclectic - almost random - mix of non-mainstream content of all stripes and orientations. At some point, we let go of the thing and entrusted the community to manage it - nobody really "owns" it anymore, and it's off on its own in terms of what it does, who runs it, where it goes. And at last count it's got about 17,000 registered members, and has tracked more than 400 terabytes worth of data. It's completely, fundamentally non-profit and always has been - totally self-organized, out of nowhere essentially. It "made itself up as it went along," with nobody outside telling it what to do. It has created a trove of content that people can mash-up, put together, share, discuss. Most of it... who knows where it comes from? Very little was commercially-produced, and only a minority is explicitly sexual in nature.

Some of it, it caters to things like bizarre Japanese subgenres involving nursing outfits, latex, cleaning supplies, and... ok, you get the idea. Really novel, interesting stuff. There's names for it - names for entire genres of content like this that basically don't exist in North America. It's all to be found on trackers like that one - the one we let go of and that's grown into its own community, its own form of organization. Folks obsessively organize stuff like this, comment on it, categorize it. Clearly, this is all way past the mainstream "adult industry" world - the more vibrant parts of our culture are very much past the entire "make more plain-vanilla boy/girl porn" concept, and onto new and often-uncharted terrain...

Personally, I think it's just... [laughter].

The scary thing is, if you go to mainstream adult industry conferences and whatnot, they're just deer-in-the-headlights when it comes to stuff like this. They really are - totally clueless.

Violet: They are. Absolutely.

Douglas: They have no idea what's going on, and they're just sitting there going: "we need to put more DRM on our 'product' so people don't 'steal' it from us." My response is usually along the lines of: "*nobody* is stealing your stuff, folks - indeed, not many people actually want it enough to bother stealing it, let alone paying for it!" I mean it - that's the reality out there, in the market. The mainstream folks in the industry are churning out stuff that people don't want - the producers who *are* making stuff people want, they can't make it fast enough and their customers are more than happy to pay for it! Even if they try to give it away at no-cost, loyal and appreciative customers will *insist* on paying for the specialized, well-done content as a matter of pride, and to support the process overall. But, if companies are making stuff that people don't really want, then there's no magical technology or DRM system that is going to force them to pay for it. That should be pretty self-evident, eh?

Unfortunately, folks in the mainstream adult industry - and this includes folks up to the senior-management level, consultants and industry bigwigs - they've gone glassy-eyed. There's all this open space in the market, all this opportunity! It's a really fascinating time, nowadays, because as I see it the industry incumbents have essentially conceded all the vibrant and fast-growing areas in the market, and they're wide-open to whomever is willing and eager to experiment.

I think the next ten years, in the "adult industry," are going to be spectacular - a real opportunity to turn away from mass-produced, disrespectful, lowest-common-denominator "porn" and move towards more interesting, more engaging, more specialized, and much more respectful ways to celebrate sexual intimacy on all levels. It's a good time to be in the industry - or joining the industry from the outside, as far as I see it.

Violet: It is wide open. It is wide open and also not to plug another conference, but Video Camp SF on October 16. I will be interviewing as the keynote Eon McKai, who is Vivid's main director, about the death of DVD. He is sitting there watching his product to get DRMed up the ass while he is watching everyone share it on BitTorrent. He announces his videos on his Twitter stream: He will be like here is the BitTorrent for the film that I don't even have out yet because he just sees it all but he can't do anything in the industry. It will be an interesting talk and it will echo a lot of what you just said.

Rose: And so you mentioned things being DRMed that can't be given away. That sounds a whole lot like what everybody's dear friend Cory Doctorow talks about in the beginning of *Someone Comes to Town, Someone Leaves Town*. The protagonist mother is a washing machine. And in *Glasshouse* [by Charles Stross], the protagonist begins the book as a tank and the protagonist is *sexualized* in *Glasshouse* as a tank, and then goes on to not be a tank at other points in the book.

Douglas: One more subject I'd like to address, if I may. In the last few years, I've authored a series of articles for Xbiz Magazine (www.xbiz.com), which is one of the biggest mainstream adult industry publications. It's a publication that's really done its best, under editor Stephen Yagielowicz, to bring outsiders to the table in order to talk with them about what's going out in the bigger world, beyond the conventional adult industry. I've also presented at a few of the mainstream adult industry conferences, primarily focusing on technological issues.

And, as we've noted previously, that mainstream part of the industry just doesn't want to face reality when it comes to the DRM question; they just want to lock things down more and more, and it's a form of death spiral for the industry, going down that path. Oddly, most industry insiders don't actually want to discuss the structural realities, or argue different interpretations of the data. Instead, they just get mad... I mean, really angry. They will start yelling at anyone who suggests DRM might not be the path to success and a return to the "good old days." When I'm faced with those sorts of attacks, I try to remind them: "look, it's not me! I am not the one; I'm the messenger - don't shoot me because I'm telling you what reality is, out there, in the real world." They'll actually accuse me - me! - of "stealing their content," which is both funny and sadly ironic. Seriously... I don't want your mainstream content. Not even a little bit; it's sort of self-evident, isn't it, given who I am (i.e. Old Blood)? They're in a death spiral, and it seems fueled in equal parts by anger and misunderstanding - as is so often the case when good people do dumb things.

Anyway, beyond that issue of DRM, I've written a number of articles for Xbiz on technological subjects. One of them, (in)famously entitled "The Fantastic Future of Porn," speaks to the crossover point where GCI-enabled technologies fully cross over the so-called "uncanny valley" and come to create mediates escapes. In this kind of scenario, customers are no longer buying (or not buying, as is the case more and more) "movies," not even CGI-rendered movies (which is a bit of a dead-end, in my view). Rather, customers are buying *access* to CGI-painted worlds, worlds that can be molded into anything the imagination can conceivably envision.

These kinds of business models, fully-rendered, multi-user, distributed CGI 'scapes - they are going to absolutely eat CPU, i.e. processing power. Right now, it's possible to more or less shut down a whole country's electricity grid, with process usage, just trying to render stuff like the Avatar environment. But, these tools to create these dynamically-rendered, natively three-dimensional, uber-lifelike 'scapes already exist... and people are going to trip over themselves to pay for access to such 'scapes when physical intimacy is fully integrated into them. They'll pay serious money for even 15 minutes of fully-rendered, fully-interactive, non-linear escape into their special Japanese table porn world - a world with different gravity relationships so they can sidle up walls easily, or whatever... point is, the more customized and unique and un-"realistic" these worlds are - and the more they cater to hyper-specific sexual preferences and tastes and orientations, the more people will rush to pay for access to them.

The tools to make such worlds already exist; it's just that nobody's actually *doing* it yet, in terms of making these worlds and selling access to them. I'm not talking warmed-over, third-rate graphical interfaces slapped on top of what are basically the descendants of MUDs; I'm talking the bleeding-edge graphics capacities that are sitting idle on just about every PC being sold anywhere in the world today. But, sooner or later, *someone* is going to do this - and do it right. It's easily a multi-billion dollar industry: just cross the porn industry, the gaming industry, and the Hollywood CGI-laden blockbuster industry... there's no way that the resulting intersections, in terms of new market potential, don't add up to billions. No way.

When I first started talking and writing about such things, in the mainstream adult industry, they just didn't want to hear it - too "far away," to different from the old successful models, the "good old days." They just don't want to hear it, still. It's like convincing buggy-whip manufacturers that making better buggy-whips is *not* the future of success; that whips are not needed for the new "horseless carriages," but on the flipside there's all but unlimited potential in making cars, and car accessories, and car repairs, and so on. I've gone further, and challenged the adult industry to start thinking hard about the ethical issues that will inevitably arise from these explicit-potential, fantasy 'scapes: what happens when these 'scapes are populated by (100% virtual, non-physical) "children?" This kind of stuff is going to happen - that and more (virtual "snuff porn," anyone?) - and the mainstream industry is, once again, deer in the headlights on how to deal with such concepts. I've seen it myself, in terms of being of a non-mainstream sexual orientation: the mainstream industry is still vestigially ambivalent about gay sexuality, much less anything "weirder" - they tend to ignore those of us on the fringes, cast us as outsiders, and leave us to sink or swim on our own. I suppose the idea is that it's "safer" to exclude us from the industry, and pretend we don't exist - same with the question of virtual ethics. But, I argue, these kinds of things *are* the future of the adult industry - the only future where there's actual margin, actual paying customers, and actual innovation. Just making more Southern-California porn with the same interchangeable "porn stars" is not doing it, business-wise, and the good old days are NOT coming back... not ever. It's forward into a much more diverse, interesting, challenging - and, yes, ethically more complex - world.

The adult industry world doesn't want to talk about these issues, embrace these issues, debate these issues. It's a shame. But... *somebody* has to start taking about these things, because the market is already leading developments into these directions, with no ethical guidance or consideration whatsoever, by itself. The market is already leading itself, and I think it behooves any collection of compassionate and creative folks, within or nearby to the adult industry, to take these issues seriously: to face them directly, to ask challenging questions, to seek diverse perspectives and creative analyses.

I just don't see much evidence that the conventional, mainstream adult industry is going to take this leadership role, unfortunately - they're just pulling further and further back into that California-centric market, and market philosophy - i.e. Southern California is the center of the universe, and if it doesn't sell or "make sense" there, then it's irrelevant to the industry - and they're more or less hoping some magical outside force will "bail them out," along the lines of the banking industry, or whatnot. It's not going to happen - the market will reward those who provide services that customers really desire, in creative and exciting and novel ways. If the "industry" won't provide that, and the incumbents continue to be more focused on dejected naval-gazing than forward-looking creativity, then others from *outside* the mainstream industry will step in and take over the lead. There's already signs this is happening, and my prediction is that this trend will only accelerate, exponentially; mark my words... there's a fantastic future for "porn" - whether the "porn business" is the one to provide it, or not.

Johannes: There is one... at Arse Elektronika 2007, Aaron Muszalski gave a talk about rented porns and stuff like that. And he was telling the story about the American remake of *Lolita* where the actress of the main character "Lolita" was under 16. So it was not possible to film any nude scenes with her so they copy-pasted pretty much like the breasts of some 18-year-old girl on to her. So, they could actually create like a nude scene and stuff like that. So, there's lot of things going on there in the realm of like rented porn. If you create render-porn like "render child porn"... is this child pornography or not? So necrophilia, whatever. The sky is the limit I see or hell whatever. But, that actually would bring me to one question because we are talking about the future of sexuality and porn here.

So, there is... Maybe it is a little bit strong but I try to emphasize it OK. So, there is the one side of the globe that is like the Western liberal world, where they would be... There's stuff happening we couldn't imagine just like there is... Like a liberal society and if you can pay for it, it will happen, yeah. And there is this other part of the globe where like women are still in the certain way forced to wear like head scarves and stuffs like that.

So, if you are talking about the future of sex, so how does that like... How does this like acceleration or deceleration of like cultures going to this hole.

Audience Member: A lot of pornos.

Tatiana: But I would say that in US prostitution is still considered illegal. So, there seems to be something to do about it and the right of the sex workers is a big issue here. The Western world is not always so liberal as we imagine...

Saul: I think the culture divide there isn't as sharp as you are making out. Specifically looking at the Muslim world, there's a much clearer division between what is appropriate in the public sphere and what is appropriate in private. The society is much more diverse than we see because we only see the public sphere. They're not going to let TV news crews into their houses.

Johannes: It's interesting, especially in Iran. Iran is the country with the most cases of female plastic surgery in the world... like doing lips and noses and breasts. It's an interesting thing to reflect, that they are, in a certain way, hindered to show themselves, but behind the walls and their clothes they are doing it to their body...

Talking about the future, talking about body modification, do you see it in different cultures in different ways? Is there a diversion between two worlds? The liberal West and the crazy frontier?

Carol: But you know, it's almost not like a continuum. It's almost like a pixilation, where every culture's got highly sexual people, sexual adventurers, whether or not they're open about it, whether or not they *can* be open about it. I'm on the judging board for the "Sex-Positive Journalism Awards", and one of the pieces that won last year was about sex parties in Iran, home-based sex parties. It was an enormously fascinating article, and I just bought a book about Syrian lingerie for the Center for Sex and Culture library last night. See? It's all coming together.

But, it seems to me that this notion that some cultures reinforce liberal exploration... That's so philosophical, as far as it goes, but there are plenty of people who aren't coming along with that notion. There's no place more obvious than the United States right now.

Rose: The Syrian lingerie book, it is an astonishing book.

Carol: It's cool.

Rose: I would absolutely recommend that you [speaks to audience] look that up in a bookstore. It's worth actually holding and looking through. The thing I want to follow up on what you're saying here, Carol, is that here in the United States, I think that it is very, very easy for us to lose track of in a room like the one we are in that we can't talk about a monoculture in referring to a country. It's ridiculous to talk about Iran, or Muslims, or even the United States.

My rights as someone who's kinky, or poly, or perverted, or genderqueer, or whatever, that I feel perfectly free expressing here, just don't exist in the same way if I were to show up in Louisiana. I'd have to express myself very differently if I wanted to feel safe. So there are issues of safety, but then also issues of propriety. How much do I want to be in someone else's face with my expression of self?

So, our cultural norms across the country are very diverse. That was one point I wanted to make, but then another... Carol is very fertile thought material for me, and always has been. Has been for such a long time. I'm a Carol groupie.

Carol: Aww...

Rose: Bit of nerves being so close to her. Carol said earlier a lot of really amazing stuff about women looking at male-oriented porn, porn images generated with men in mind. I'm guessing that the first thought you had there was porn which had men and women fucking in it.

Carol: But, one thing that has been true for a very long time is that women have been interested in gay male porn. So when we think about slash, or when we think just about women getting access to gay male porn, there's an interest there in looking at gay male porn for the same reason that men like to look at lesbian porn. There are no men in lesbian porn, and so men get to just look at women fucking. [In gay male porn] there are no women, and it's there for me to just look at men fucking.

Rose: That's a way of choosing to look at porn which was not intended for you, but which satisfies a desire that you have, right? Those are things that have been going on for as long as there's porn. You have to make choices about what porn you consume, even if the porn wasn't made for you.

The one thing I would add to that is the other thing that you're doing, and able to do when you're looking at two people of supposedly the same gender, or as far as you can tell by looking at them on a screen the same gender, because you don't necessarily know for sure that that's how they are going to identify is that you can identify yourself with either or both of them in a way that is a little harder - not impossible, certainly - but harder if you're looking at heterosexual porn. I think that's another of the reasons that it's attractive, that same sex porn is attractive. This notion that I came up with, under feminism, that the reason that men wanted to watch girl/girl porn, as it is always talked about in the industry, is that they want to imagine themselves to be the only man in room. It is sometimes, in fact, that he wants to imagine himself female in the room with them, or wants to imagine himself with his male identity, but not having the kind of male oriented, or male-categorized sex that men are taught to have.

You've got to break it down so carefully and listen to individual people's thoughts about this kind of stuff. I'm glad you said that.

Eleanor: I wanted to bring it back and react to what you had to say, Johannes, about Iran and the "Two Worlds" we live in – because I was the one that was vigorously shaking their head over here.

Johannes: Bring it on!

Eleanor: That's OK. What I thought was really interesting about what you said, and then Carol touched on it, talking about the pixilation of sexuality, and then you touched on it Rose, talking about sexuality in context. In terms of what you said, with sex in different cultures, and thinking about sex in the future is that you touched on what a lot of people have when they imagine sexual freedom and the lack of it in other cultures.

There's a feeling of overall sexual xenophobia that I think is starting to emerge in a lot of the conversation that I'm hearing, and especially with... I forget the gentleman's name that was sitting here earlier. He was talking about being an XBIZ writer, and talking about the future, sort of Second Life-style worlds and how anything can go. People are just going to act crazy, and there's going to be worlds full of children getting raped and stuff like that.

This is the sort of sexual hysteria that I keep hearing over and over again, that's a sort of underlying fear about what sex in the future is like. It's the same sort of xenophobia that you were talking about.

The answer is really in the contextualizing that both you two were doing... You three, actually. I was looking at Kyle in the back a little bit, because he used to work at Linden Labs. I think that, not to beat a dead horse with Second Life, however there are people that work there that I work with at San Francisco Sex Information. We worked the hotlines together for years. It's a very, actually, sex positive organization in the background.

It's a self-rectifying culture, in many ways. It's self-moderated. It's a very interesting example of when a culture is allowed to do what it does within the parameters of anything can go and seeing what happens. People haven't gone crazy, and shot up their families at home. They haven't gone off and fucked a million sheep. They haven't turned into Phillip Garridos who are already that way and probably don't have access to things like Second Life or whatever, to go and express their sexuality.

It's this hysterical idea that once anything can go, sex in the future... Once we can fuck anything or disappear into our own worlds. Once we have a chip implant where we have orgasms on tap, are we just going to turn into these drooling blobs in the corner in the dark, just sitting there pressing the button like the rat waiting for the cheese to fall out until we're so fat our stomachs explode?

It's like, "I don't think so," because we're human and each culture has its context. Each facet of sexual expression has its context. There's always some part of culture that's going to moderate that and help shape that so that it's in its proper context one way or the other. So, don't fear the future.

Rose: To go along with what you are saying, technology isn't new. The idea of technology being the thing that we don't know about, the thing that's going to happen, then is going to produce this pedophilia or whatever...

Audience Member: Fear. Fear.

Rose: ... insert fear into this blank. Jason Scott gave an amazing talk about technology in relationship to porn and gave us examples of typewriter porn and porn being transmitted over Ham Radio. We have at the CSC library, at the Kinsey Institute Library, we have examples of porn collected over the entire course of human events. When we look at that porn, we do not see a high percentage of pedophiliac porn. Right? There's no reason to assume that as we adopt new technologies what we, as human beings, will want to suddenly change or at least I don't think so. I agree with you completely.

Johannes: Go ahead, if you've got something to respond to that.

Saul: So, one of the things that we have been talking about is people using these technologies to get exactly what they want, being able to customize exactly what they desire, being able to use these technologies to reach out and find other people. That's one of the cases where we are already seeing technology changing society a lot, and it's clearly accelerating. Look at the community in this room. Look at the talk, earlier today, on the Internet enabling polyamory.

These are good things, for the most part, but it is also driving this kind of siloed culture shift: all of these cultures may be overlaid in physical space, but they are not really communicating.

There isn't that kind of spatially related cultural bond. I come to San Francisco. I meet the people who are part of a very specific subculture. There are however many million other people in the Bay Area. As far as I can tell they don't exist. I don't interact with them in any meaningful way. It is becoming increasingly easy to do that. I'm not saying that this is necessarily a bad thing, but that's a really big shift in terms of how people interact with culture. I think it's going to be really interesting to see what that does roll forward 50 years, rolled forward a hundred years.

Johannes: Perverted futures?

Saul: Yeah. I kind of wanted to ask a stupid question which relates to that really, which is what is porn for? What are we using it for? Because all of the experiments and the extrapolations on pornography and autoerotic imagery, and constructions or sexual augmentation, in the most part, has been talked about in relation to masturbation of various kinds. You can call it what you want, but it just seems to be generating a lot of wanking. I'm interested in pornography in the same way that I'm interested in science fiction. I didn't like Douglas Spink's talk and his positioning of these science fiction writers on this two dimensional grid, because I think that what science fiction does is provide us with a mindset or a position from which to look at innovation or society or technological development that's really at right angles to the current market trajectories of technologies.

What's exciting to me about pornography, especially about non-human porn or non-human sexual devices and erotica is that it's this whole skill set or this whole set of attitudes developed within this community: that could completely transform the way that people approach all kinds of activities.

Academic papers for example. Some people could start feeling aroused by the particular phrasing of their academic work. Then, develop the ability to know that 'OK. I'm doing something right here, because I've got a hard on' or whatever... This could expand sex education radically, not just showing people how to have safer sex or liberate their personal desires, but actually transform the way they live their lives and imagine their futures.

Johannes: Maybe that's another component that we need to start adding to sex education is... Like sexual interpretation, how to interact with someone who has radically different sex than you in whatever dimension, so that kind there's a way to combat that sort of xenophobia and kind of just as a, "Here's how to understand someone who gets off really differently than you. They're not that weird."

Whether it's going from the mainstream to what we think of the fringes or maybe us out here on the fringes saying, "Oh, yeah. OK. Those quote vanilla people, yeah, I can sort of... I can get that. I can understand that they're not that weird."

Tatiana: I also wanted to say something. As I said the real challenge is how to make porn close to our daily life. I think that this should be our main objective. There are different possibilities and different strategies to achieve this. It's really not a matter if the technologies we use for porn are new or not. I mean, you could actually create porn using Second Life or make ASCII porn, as we saw yesterday. What is important of using technology is the possibility we have to share experiences and get closer to topics we would like to experiment with - like porn - and to make them part of the daily life. You say that the sex machine should be in a museum. I think it is a great starting point so everybody can go there and fuck the machine; or it would be even better if in the future the

sex machine will be at any street corner! That should be the mission of the next Arse Elektronika, I think. [laughter]

Johannes: Let me get a phone number here of the Museum of Modern Art here, and we'll do that next year. [laughter]

Eleanor: Just to follow up on the notion of learning how to relate to people, who have a very different sexual practice than one's self. I mean this is something that I think is what brings a lot of people into the kinky or the poly communities is that they are in a relationship with someone who might have a very different practice, and they don't want to lead the relationship.

So, then, they have to learn how to deal with that. So, I think that is actually something that is already getting discussed quite a bit in an informal way. It probably would be helpful to have more organized kind of education and to make it more OK.

Because one thing that seems like it's becoming increasingly clear is that there are many, many people in relationships, where they're pretty unhappy, because they are not really compatible in some way.

That's not really recognized as a serious and legitimate issue. If you're not happily compatible in some kind of heterosexual monogamous relationship, well, somebody is a failure there, rather than it just being, "Oh, this is just something that happens."

People don't learn to find out about it ahead of time, so that they can decide how to negotiate that as part of the normal process of a relationship.

Johannes: Vanilla people need checklists too.

Eleanor: Yeah. [laughter]

Carol: That's for sure. Sometimes even more. I really want to say that I think that, speaking from the radical sexology perspective, there is already a profound degree of knowledge about this stuff in community-inflected and -influenced sex education. When you started to speak, what I wanted to say is, "What sex education?" We have all taken sex education into our hands... I mean, talk about hacking. We have had to hack the culture around sexuality. This is going to continue to need to happen, as long as these cultural notions about sexuality and its use stay as polarized as they are. Thank you, Catholics, who are calling me a pedophile right now. [laughter]

Many of you were here last night, when we said that. But, if any of you were not, can I just say that: Catholics are calling *anybody* pedophile. Please. Please. Please. It's just so complicated.

But, what I want to say about this is that there was just a few days ago an op-ed that I read somewhere - it was in a UK paper actually I think - about a book that has come out recently asking women why they have sex, and there were 234 reasons. And they actually admitted that there are probably a few more reasons that they hadn't gleaned.

But the notion that there are only a few reasons for somebody to have sex in the first place - with another person, much less with yourself - I think that starts to get at what you are talking about, Saul. There are so many different directions to come from to get to what might seem like the same place, but then if you come from radically different enough directions, it is not the same place at all. It is different. It is the same act, but with two different reasons for doing it, it has two different meanings - and we can all think of at least three relationships that we've been in that broke up for that exact reason, because nobody taught us to have that discussion.

And because pornography has always been sought out as a source of sex education even though sometimes, it does that in a phenomenally shitty way. Sometimes better than that, but sometimes in just a phenomenally shitty way. Although I blame the pornography of Andrew Blake, where the woman stood and put her leg all the way up in the air to get fucked in the early 90s, for the yoga studio explosion. Personally, I am just telling you right now that is what I think happened.

So who knows what is going to happen next? Who knows what the next pornography is going to unleash upon the cultures? I think it is really important to say no matter what we talk about, if we call it pornography, somebody is going to reach out to it to get inspiration, support, permission - to see "the thing itself." That is always going to happen unless the culture has replaced pornography with something else.

And so far, many of our cultures have not done it very successfully.

Rose: So we are talking here about pornography as sexual education and then we are talking also about sexual education more broadly as a way of improving relationships and relationship styles. We were also talking about sexual education as getting to the heart of what education is. I found it very exciting. Speaking as an academic but as someone who has very ambivalent relationship to traditional academia, one of the reasons I have that ambivalent relationship is because I am so passionate about the idea of teaching. And there is a sort of a... You could call it a dirty secret, which is that in the teaching act, you are sort of channeling that attraction of the teacher and the student of the relationship between the people that are doing the learning and the teaching, right?

And then what you [Saul] were then also referring to is that excitement which we channel. We intellectualize and don't call real excitement. We don't call it physical, sexualized excitement; we say it's excitement of the mind. We don't say like "oh, that makes me wet." We don't say "that makes me hard." We don't say "that turns me on." And so, you [Saul] were talking then about the kind of, say, sex machines that are made and the electronic tinkering that goes on.

And you refer to it as being legitimate science education and that the sort of people who are willing to say this turns me on and I want to do the tinkering are not necessarily [sexual]... Although sometimes they are. Our top-notch engineers, and I think that was what you are trying to get at... That if we could be perhaps more open about our sexual motivation, our fullest drives in whichever pursuit we are in then the sky is the limit.

That is what I think Tatiana was talking about too, that if our pornographic sort of interest, if our sexual imagination can be part of our everyday imagination, the...

Tatiana: Yes...

Saul: Totally. No, that is *exactly* what I was grasping at because it seemed to be legitimizing education or legitimizing innovation in all kinds of fields. I love the idea of the fuckable art show. I think that is a great idea. Similarly, how could you have sex with the Hadron Collider? There are all these kinds of tantalizing opportunities once you legitimize all these practices and career options. It's like the issue politics revolution that happened in San Francisco and spread to the rest of the world that has transformed people's relationship with themselves and their bodies. It could go so much further. In fact these communities could take that kind of libidinal energy and distribute it to other communities of creative people.

Johannes: I had sex with an overhead projector three years ago. Actually, I had sex with an overhead projector because I wanted to prove its pornographic possibilities. I ejaculated on the projector... and on YouPorn it has a 3.75 rating out of five. So it is not bad. [laughter]
The problem is: I didn't like having sex with the overhead projector. I did it because I had to do it. Only once.

Audience Member: Virgin!

Johannes: Yes, yes, my first time. That is the point. That is what I am saying. I need to like it, kinda force me to like it. Perverted, but necessary. I need to develop the basic drive, I guess. [laughter] So, I know I promised to have sex with another overhead projector, and maybe some table, chair, trash can, whatever. I promise until next year, I will have sexual intercourse with at least 40 or 45 different objects. I promise.

Eleanor: That is good because if you are goal oriented, your first time would be disappointing.

Saul: I will tweet it. I will tweet it.

Eleanor: Wow. And so how to have sex with the super collider? Really quickly. I want the fucking machines toaster sort of like bicycle. Does anyone remember? Was it a tandem bicycle? Does it move like a bicycle?

Audience Member: I have done it in a seat!

Eleanor: Yeah. If it could move like a tandem bicycle. I know the super collider used the bicycle to get around the super collider because it is like... It is long. Because it has to be long because the atoms have to go far so you could sort of have some kind of tandem bike or single bike or...

Audience Member: I can fuck the super collider.

Johannes: I think that all of you should have sex with an object in the future, OK? And I think we should find a twitter hashtag.

Rose: And maybe it should be sponsored by IKEA.

Audience Member: Table or something.

Johannes: A hashtag, like it was a secret codes. A very discreet code.

Rose: How about hashtag dollar sign table (*#$table*) because table is standing in for whichever object. Is that too geeky?

Audience Member: Yeah.

Tatiana: We could do an open contest and see which proposal...

Johannes: To find a hashtag? Come on! [laughter]

Johannes: Let's stop that Social Web stuff and let's really talk about the future. So my question is: If you could invent something right now or if you could change something on your body or on somebody else's body, what would you do? What kind of new gadget or new body modification?

Tatiana: I would say, I don't know what is the object but it has to be something that make me laugh when I have sex...I think it should be ironic and a pleasure, that is also kind of a game, something hilarious. I don't know. I would think about that. [laughs]

Eleanor: Is anybody else ready with their thoughts?

Johannes: Tough question, I know.

Rose: Oh! There's frowning. There's frowning. So I've been very interested that disability has come up a few times over the course of the conference. I would say, I suppose, I'm sort of invisibly disabled. I have rheumatoid arthritis. And I do use a cane, but you can't see right now that I use a cane. So, what would I do? I would not have arthritis but whatever.

But as we start, I was really struck by some things Ella was saying and, of course, some things Annalee Newitz was saying about, you know, what sort of prosthetic things we might have or if we could, you know, pick our legs at will or our arms at will like, "Oh! This hand is nice or I know how to use this hand, but I could pick the other hand."

Johannes: OK.

Eleanor: I have a set of hands.

Saul: Well, I've been, I can kind of cheat on this question because I've been spending the past couple of weeks thinking about what I would change.

Carol: You swotted up.

Eleanor: And I think my favorite of all of the fictions I threw out this morning was skin grafts that are based on cephalopod chromatiphores that you can...

Johannes: That was the best phrase of the conference.

Eleanor: ... that changes organically in response to your emotion. So you have this completely new affective channel. You have body language, but you also,

well, you turn pink when you're embarrassed and you turn black when you're angry and you turn spotted when you're horny. So that's what I'd like. [laughter]

Carol: Well, my old fantasy is to have a prehensile tail. I think there are an awful lot of sexual uses for a prehensile tail. Excellent! But I also want to repeat to you this great idea that I have adopted it since I heard it - it was the winner of the "Good Vibes Millennium Sex Toys You Wish Existed" contest, and somebody reached all the way out in the middle of the millennium for this idea, I think. But I'm also sure somebody's working on this now. In fact, I'd totally love to hear whether you people think the technology for this is close versus far: to be able to record your own sexual experiences and play them back later, so re-experience them. And, ideally, to be able to allow someone else to plug in, however that's going to be done, and experience the same sexual experience that you've experienced.

Rose: Oh yes. Hot!

Carol: I think it would revolutionize practically everything, personally.

Johannes: That's like the movie "Brainstorm" from the early 1980s, starring Christopher Walken. They record everything. Death, sex. Someone's brain gets fried because he makes an endless tape loop of a recorded orgasm. He experiences it for a couple of hours and turns into cabbage. Good film. You should all check it out, although the end is a little bit too United States of frickin' Spiritualism. It's somewhere on Pirate Bay, I guess. [laughter]

Saul: OK. I was reminded earlier on today, when there was some early talk about inter-species sex, of the fact that bacterial sex is now seen as one of the primary transmitters of genetic information between humans, not necessarily sexual reproduction, but the exchange of mitochondrial DNA through infection. And I'd love to be able to feel that. To develop some new kind of nervous tissues or senses so that all the things that were constantly having sex within me, like when I caught a cold or when I got some sort of horrible disease, that I'd actually be able to experience pleasurable feelings or some kind of feelings other than just getting a blocked nose. [laughter]

Johannes: Wow!

Violet: Well, I am nothing if not a practical girl. And taking into consideration what everyone has said, first Carol read my mind. And I've thought of a couple of things while everyone's been talking. One is that, the first thing I thought of was the ability to record the porn that I see with my eyes when I go to these things that I go to, when I go on set to people's porn films or when I visit the *kink.com* castle. I see amazing things that I can't really record or talk about. Sometimes they let me take photos, but if I could record that and have some mental space that I go to, that's a private space that I share with my community, where I can share what I've see, that would be awesome. And then the other practical thing - actually, there's three practical tips here - so the other thing, I was so pissed off when they stopped the research on Viagra for women because it totally fucking works for girls. And they've discontinued the research saying that, "Women are more interested in romantic fantasy than physiological stimulation."
And I was like, "Fuck you and your mother. I have tried Viagra." I have this great gay doctor and I was like, "What's that stuff I can use?" And he's like, "Honey, take some home." And I did. And the problem was that with Viagra and Cialis and the other things that I've tried, I have allergies. I'm a geek. I use an inhaler sometimes. It makes my nose so stuffy that it's just not fun.
But it works. It's like, say if you can think of your arousal in three stages, skip one. You're halfway through two, and that's what it feels like for girls because we have the same erectile tissue that it acts on in men. So, it works great, stuffs up my nose. Fucking finish the research and get me a pill I can use because I want the female sex pill.

Audience Member: Viagra does have different effects on different men!

Violet: I agree. It doesn't work on all men either.

Johannes: Oh, come here. Tell us. Tell us.

Violet: That is what was in the press releases that they put out.

Audience Member: Women's arousal systems are supposedly too complicated.

Violet: Oh, god! Nice!

Audience Member: And in one sense it's true because not every woman keys into what you're keying into when the blood starts flowing down there. You feel it. For you, that's a signal for arousal. It's not a signal for every woman, and that's where it gets complicated. So for some women, pink Viagra works. For other women, something else, a different fill up. That's sort of my specialty.

Rose: Yeah. And it's the same situation with men as well.

Audience Member: I agree.

Violet: Yeah. And then the third thing I wanted to add was actually something about the title of Hypercrotch and Nanobot. Which is if your hypercrotch does contract nanobots... [laughter]

Violet: ... it's a one to three day lapse in time when they start to manifest. You get it from skin to skin contact. Go see the rubber nurse at *kink.com*. She has a salve just for you. [laughter and applause]

Johannes: So I have many things that I would like to have. Most of them are so perverted that I will not talk about it here. [laughter]

Saul: Where else can you talk about it?

Johannes: I think I have to invent a different conference to talk about that. But, OK. There's at least one thing I really would like to have. And it's derived from an idea out of Iain Banks novel, *Matter*. It came out last year. Iain Banks is like the godfather of this conference, because I incorporated some of this ideas in the call for Arse Elektronika 2009. I was talking about his fictitious group civilization called "Culture" that's pretty much like an open, anarchist, socialist civilization, and they're pretty open about their sex. They all are changing their bodies all the time.

And in the last one, *Matter*, there is this female special agent, and she has a small drone that protects her and supports her. It's her friend and her companion. At one point, the drone has to hide, or in a certain way, disguise itself, and it disguises itself as a dildo.

I kind of like the idea of having a super intelligent dildo or butt plug. I want to have a good conversation with my butt plug. A really good one. [laughter]

The rest is concerning stuff like injecting materials into my bloodstream, but... Talk about that later.

So, yeah. Any questions from the audience before we close the panel?

Audience Member: Actually, it's not a question. It's an *offer*. I run an erotic literary salon, so I'm basically working with people like you were mentioning. We're a select group here. We're used to hearing all the words that we're hearing, and all of the ideas. I'd say the majority of people that come to my salon for the first time are venturing into an area that they have no idea where they're going. Most of the time they don't tell anybody they're going there, and then, eventually, maybe they'll even bring in a friend or a family member.

One person has been writing a novel for five years, and he said if his wife ever knew, she'd divorce him immediately. Some people, this is the only place they can come comfortably to talk about it. So, I'm offering them, as a sample group for anyone who would like to run things past these "normal folks" out of Philadelphia.

Saul: People spend a lot of money on lasering and plucking and waxing. How about having programmable pube patterns and you can specify preference for your partner. It can come up, like a Brazilian, or really bushy, or whatever. That would just save a lot of time and pain. You could even change the bore or the thickness, whether you like it rough and scratchy or soft and feathery, carry like a whole portfolio of pube patterns with you.

Audience Member: I just want to say, before you leave... I really like your fantasy about how you can get sexually aroused by every activity that's going on inside of your body, whether it's an infection, or cells dividing, or something like that. There're probably some psychedelic drugs out there that could be slightly modified to do that right now, LSD based or something like that. Because then, all of a sudden, you really become aware of your entire body, and any one mitochondrial action will just make you go, "Whoa!"

Carol: Yeah.

Johannes: OK, so any final statements? No? Okay! So, a big applause for the panel, please. [applause]

Rainer Prohaska

FUCK #XXX

Under the title "FUCK #XXX" artist Rainer Prohaska creates sculptures, pictures and videos with pornographic content. The initial elements are mostly objects of everyday-use, or derive from pornographic areas themselves.

Rainer Prohaska works on artistic concepts, which take on common processes, and transforms and presents them as a modified reality in fine art projects and performances. Via minimal interventions everyday actions are altered to original procedures. Cooking, constructing, mobility or sexuality becomes art.

At the same time the artist experiments with modular temporary sculptures. He uses mobile and household objects or elements from construction sites, for large-scale architectural interventions in public space. Toy kits like Lego and Matador inspire the assembly methods of such sculptures and objects. In these experiments the "Performative Act of Constructing" and the "Effects of the Public Space" on this process play a crucial role. How can mundane objects be assembled to creations that, at first sight, look common, but irritate the audience in the end? How much space has jest and humor in art?

"FUCK #XXX" takes up and combines those two working methods. Sexuality and pornography, basic human topics, and their practical outputs (masturbation, for instance) are raised to creative actions. Items, normally used for banal tasks, are used like toy kits, get marginally changed via applications of other materials with flexible connections like cable straps and ratchet tie downs, and in this way are transformed to sculptural stimulation-devices. Their pornographic aspect is visibly recognizable, and can be experienced while using them practically.

"FUCK #1" is the first group of those objects made of re-designed household appliances and tools, like electric engravers, wooden spoons, food processors, pliers, etc. These objects form a sculpture park, which offers suggestions for imitations and also for advancements of such devices. Additionally, Prohaska plans

to form a DIY Porn Tools Community in connection with "FUCK #1".

This project should touch the audience on many levels, by pairing unsuspicious materials with pornographic content in a surprising and in the same time confronting way.

Text by Eva Grumeth

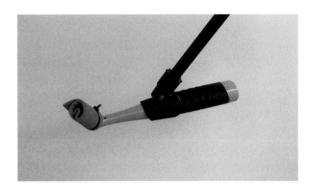

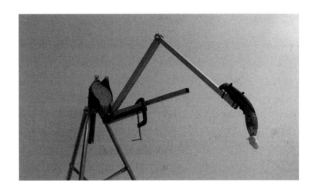

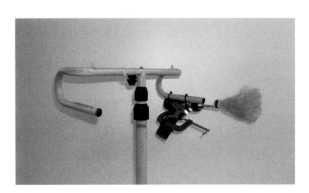

Thomas Ballhausen

WHATEVER HAPPENED TO THE IDYLL OF THE KINGS?
NARRATING DESIRE, REPULSION AND PHOTOGRAPHY THEORY

I don't think that it's
Gonna rain again today,
There's a devil at your side,
But an angel on the way

Someone hit the light,
Cause there's more here to be seen,
When you caught my eye,
I saw everywhere I'd been
And wanna go to...

You came on your own,
And that's how you'll leave,
With hope in your hands,
And air to breathe

Editors: An End Has a Start

He is already up, before the alarm clock wakes him as usual. His anxious glance wanders over the darkness on the ceiling. Although he has slept just a few hours, he is, very different from normal, alarmingly awake. Slowly and silently he gets up and slides off the bed, switching off the alarm clock as he walks past it and glances out the window while drawing the curtains. The recent storm has calmed down, no rain. He slips on the prepared clothes that he had arranged in the correct order on the backrest of his chair the day before; the mistrust he has for the accuracy of the weather-forecast which has promised sunshine and shining temperatures, supports him. He grabs the already packed bag and walks out of the room without making the bed or opening the curtains entirely, as if he wanted to arise the impression to others that the resident of this untidy robber's den just got up for a moment, according to the early hour, to get back to the bedroom at any swift moment or return from the toilet to continue sleeping on this Sunday morning. Being the only one rendered by this razzle-dazzle for the moment, he stands in the corridor of his apartment already, choosing a thick black jacket and a pair of shabby looking but rather indestructible sneakers. He grabs his key and assures with a glimpse in the living room that his doing was not monitored, meaning that his flat-mates are still sleeping and not secretly watching from the enormous sofa. The door shuts silently, just like he was hoping. Swiftly, although not in a hurry, he flies down the stairs and leaves the building. The air of this early morning is cool and pleasurable; a soft wind wipes the remaining fatigue off his face that was again produced by the weight of the bag resting on his shoulder. According to his plan, he walks part of his way to the appointed venue. There are almost no cars moving, though sometimes taxis drive pass him on his way. He walks straight ahead for some time and then takes a right, following a steep upward alley that brings him to the train station. One of the houses on the lower side of the street has white, metal balconies. But instead of other early birds, crows sit there, in pairs, watching him with the security and serenity, like tenants that have paid their rent months ahead. They indeed turn their heads synchronically – he can see as he, without slowing down, turns towards them – to watch him until he finally disappears behind the next corner from their range of vision and he notices today's newspaper headline: MORE THAN JUST A BEAUTIFUL FACE, written in bold letters.

She is already standing there, downright surrounded by a huge number of bags, and she is really wearing her new trekking-shoes, with which, as announced in a phone conversation the evening before, she intended to move in carefully and slowly, not to get blisters or pressure points. The hasty welcome kiss and the absence of a hug mirror both of their tension; their aim is to lower this tension over the upcoming two days. He already feels that he is not really willing to let that happen. They have a rather trivial, shallow conversation to help the time pass while they are waiting for their friends with the car. She makes the impression to start with another phrase, but stops, the mouth already opened, gets a cleaning cloth and starts cleaning her shades. Wordlessly she passes it on to him, a gesture of evaporated trust. He takes the cleaning cloth, takes off his glasses and for a moment enjoys the fuzziness that now covers the surroundings.

Just as he wanted to get away from this awkward situation with some lame excuse like getting coffee or a newspaper from one of the shops in the train station, the expected others arrive with a spacious car and, after getting out of the car, immediately start to flock around in their intrusive way and with plump friendliness start talking; a circumstance that is not even interrupted by the action of putting the luggage into the car, the loud statements of the obviously already discussed seating arrangements nor the uncontradicted taking of the mentioned seats. He takes one of the back seats with the feeling of betraying his own resolutions again. The alleged intimacy of the reputed friends is continued by the template-roles everybody thinks he/she has to play like in the daily soaps: there is the jester, the shy one, the idiot, the beautiful princess, the sensitive hermit and so on and so forth. He asks himself which role he would have, while he leans back trying to observe everything without taking any part in it. He likes himself when he enjoys thinking of himself as being the silent and underestimated one, to hide his actual role of a smiling and well-educated axe-murderer. He wastes a few more thoughts about this construct and finally comes to the conclusion that he might be the fool, maybe even the idiot. A wave of sentiment catches him in this phase of senseless conversation and shakes him to and fro like on dull and calm water. Just when the front passenger starts talking about her new job as a test-thief and she mistakenly says test-burglar first, he listens a bit more attentively to the unusual depiction. But only a few sentences after her description she stops and falls back to the him-excluding tone of the conversation and they continue talking with each other, as if they were very close together. But with every third sentence they are completely wrong, constantly quoting movies and soaps they have not even seen themselves. Paradoxically they keep correcting each other and try to hold on to a picture of a long conspired community. He is pacified about the smattering they have about him. Even together - and that makes him proud in a humiliating and childish way – they could not reach an approximately correct nor complete picture of him. This thought which seems to him like a certainty, brings him back to the question

that bothered him upon waking, on his way to their meeting point and during the clunky salutation: what am I doing here?

The car passes by numerous construction sites, the usually flat concrete surface is broken up, boulder, dirt and machinery are in the middle of the lane. He hopes to fall back to sleep, not to continue his halfhearted involvement in the conversation. But he is neither granted sleep nor the mercy of a small crash or a severe glitch that would bring this trip to a sudden end. The next hours they follow the GPS until they finally arrive at their destination. The weather has gotten better, the church-bells do not just ring for lunchtime, but also – so it seems – to welcome them, as they park and get out of the car, outside their hotel. They unload the car and agree to meet later.

The double bedroom that the landlady shows them is surprisingly bright and from the small balcony they have a good view of the lake, on which the idyllic village is located. She starts to mechanically and precisely unpack her luggage and place it; he puts his bag on one of the broad chairs and smokes a cigarette on the balcony. While he blows the smoke in the clear air, he has to withhold his desire to ask for the departing time the day after tomorrow, and turns around towards her, pretending not having heard her when she has already called him three times. She is sitting on the bed, pats the blanket next to her and invites him to take a seat. Without showing his reluctance and his tiredness too explicitly, he looks at her and tries to discover again what had once fascinated him, yes inspired him. She looks at him long and hard, grabs his hands firmly into hers, similar to the religious teacher he could remember, that advised him to be more humble, and asks him, as if she hadn't seen him for years, how he was doing. He does not answer immediately. As he tries, she starts talking again about her assumptions about how he might be, and frees him from the annoying duty of answering. He silently waits until she is finished, does not interrupt her, but also does not complement or confirm her statements. She releases his hands as if everything was said and done with a relieving sigh and suggests lunch with the other travelers who must have had enough time by now to make themselves at home in their rooms, and with other friends that have arrived in the village. After all, she thinks aloud after a look at her watch, it is about time.

Actually he is not hungry yet, but eating will be something like an amusement, procrastination without importance. It's just one of their friends who he really does not want to meet; he fears every further conversation with him, his inevitable, always returning infiltration with flimsy myths that were spread by alleged introverted people about him. These stories wandered on whispering detours from mouth to ear, became more and more like naturally grown, and finally landed on him. He is not really mad about that, he might have reacted similarly in a comparable situation. He remembers the last conversation with the hated one very well, how he embarrassingly streaks his back-combed hair, controlling its position especially at the thinning areas. These gestures, the addition of slickness and spotlessness of man and mission, have manifested his picture as knightly-like philanderer that in the eyes of others becomes an advocate of honesty. He would have preferred his opposite to directly try and snatch his girlfriend. But this situation, this questioning and the threat of force stimulated him to make up and confess something, to shrive a picayune adventure like a cardinal delinquency, almost like a real crime. But his lethargy and his unwillingness to seek a mock have gained power in the conversation, and he could have also tried to catch a piano falling down a skyscraper and remain unwounded. Everything he did he could have done differently. But what for? The sense, he is thinking on the silently walked way to the restaurant in which they have reserved a table weeks ago, has not become clear to him, not even after the countless long relationship-conversations. At last she tried to convince him to take a common course with her on one of their rare trips a few months ago. They went to a thermal spring for the weekend, where they, surrounded by much older couples, wanted to enjoy some leisure and relaxation. Dawdling in the warm water, sleepy from the hot air and the omnipresent food, she again

came up with her desire for a decision, a decision of his towards her. She said the according sentences about a long-lasting bond, a common household and her desire for children, as if she had copied them from a guidebook and memorized them for this very situation. This collection of clichés that meant the truth to her were presented to him in already existing pieces of a language, that was only made for terms like homestead, marriage and children, as if the words were just invented for this. The significance of his love for her, which space aloof the planned shed she could take, did not become clearer to him. Furthermore it became clear to him that besides the ready-made nothing is possible together with her, and maybe never was. His refusal to give direct answers or to inflect to her ideals brought them to a point where she very righteously gave him the impression just to be something she has to bear with. She made her decision for him, and this silent decision that she actually made for both of them, is irreversible. There is just one option for her, which she already knows from her parents, of suffering through the circumstances and holding on to the fulfillment of a fixed and indissoluble plan, thoroughly transparent togetherness. The necessity of human secrets he believes is in big parts unknown to her, just like the estimation about how much love there still is in this situation and not just the avoidance of feared loneliness. Her inability not to dominate and his attitude to mainly surround himself with weaker people, this surprisingly seldom causes loud fights. It was a rather ambiguous feeling, a creeping dawn. The same way he got out of her control, she became blunter and more common day after day. Unglamorously, in his understanding she sank further back into the sea of grey masses of uninteresting people. If they finally would fail, following the resolution that he drew while opening the door to the restaurant, he would just disappear from her life and erase her from his mind with childish insistence.

After lunch the group splits up again and walks along the lake, following the promenades, until they reached the boathouses and the mooring areas of the local hotels. They stop at one of the fields and walk up until the edge of the massive wood-construction that extends the actual bank. The sun has made him sleepy; he lies on the foot-bridge and takes off his glasses. The harsh texture of the grey, sun-bleached wood that he feels through his T-shirt and on his arms cannot stop him from noticing her movements. At a slight distance, but not really out of reach, she lies down next to him, stretches her right arm and he can feel her hand on his. Her fumbling fingers get firmer and make a strong grasp, which he instinctively returns, without thinking about his action or what it could mean. As if she understood the halfhearted gesture and his insecure reaction, she gets off of him. With closed eyes, relying on sound only he assumes that she got up, stretched and went to the boathouse, where she sat down on the wooden pier. He sits up, dazzled by the sun, looks for her and finds her at the assumed spot, sitting, with her head on one of the wooden posts. Loudness is not wanted at this place, which makes an advantage for him, as he lies down again. He closes his eyes again and opens them, when she touches his shoulder with her foot. He did not hear how she got out of her clothes, which are folded accurately on the wood. Her nakedness that is hidden under her swimsuit cannot commence his attention; a feeling from saturation to weariness comes up in him as he is looking at her. He stopped dreaming of her a long time ago and still he persists stubbornly not having changed. She looks down on him, with folded arms under her breasts, her entire body shaking from anger. *I have to know*, she says, *that you can do something for me. That you can do some things just for me and really just do them for me.* She winks, her physicality now seems like a demonstration, an exposition for him. Would that be the crisis that he was secretly waiting for? Her challenging look makes him hope that she can wash him off, layer after layer, like a slipping photograph. In his imagination he would then be retouched, disappear as if he had never existed, as if she had never needed him. The precise removal would be like a lifelong effective vaccination that would make her immune to him in the long run. But nothing of that, just her stiff look and his fear of getting hit by a certain impulse of hers. But he wants to test the worst and most unbearable thing; he wants to test how to try on clothing that he had no intention of buying, like a masquerade that has proven not to be applicable. In a very cold way he still finds her beautiful, but he can no longer name what attracted him once. So he closes his eyes again and just listens to her demonstrative sigh, lying down next to him again and pressing her clammy hand against his, enjoying the sun.

When he wakes up later, she has already gone. It is late afternoon and he gets up very slowly because he has a light headache, caused by the sun. He walks the same way back and starts a walk through the village without any plans. He stops in front of a fussily renovated church and enters with a large tourist group. It is comfortably cool inside the building, and he marches determined towards the baroque altar and spontaneously throws a coin in the machine that was there. But nothing happens, none of the electronically candles are illuminated. Disappointed, he walks through the strange smelling church and buys a small candle that reminds him of a tea candle. On one of the upper rows of the sacrifice-candle holders there is still some space and he wants to light his candle on one of the already burning ones. But the wax that is dripping down his candle extinguishes another one. Afraid that every other trial might lead to a cruel slapstick comedy and another attempt would worsen things immensely, he simply puts his candle on the candleholder. Leaving behind the irreversible, lurking mini-catastrophe, he storms out of the Lord's house and after a quick thought enters the next café. He takes a table offside the tourists with their projecting gestures and the trophy-like worn local clothes which they combined with their overpriced function-wear, and orders from a well-experienced service person. The red jumper of a loud man in the adjoining room reminds him of an encounter on the train in his city a few days ago. The awkward mixture of aromas during rush hour and of the crammed wagons filled with people simply made it impossible for him to read during his train ride. Amongst the passengers was a young lady in a red dress and a fabric-jacket of the same color that stood right in front of him. As they accidentally looked at each other, smiled unsealed, it gave him one of these almost magic moments. She eyeballed him, noticed the book that he had closed and held like a bizarre shield towards his breast. As if he could watch himself, he saw how he observed her without moving noticeably, how he imagined taking her hand spontaneously. For a very short moment he was craving something that was not meat but nearness. She was wiping her face with her hand and left two black stripes on her cheek and discreetly replied his smile. Other passengers had noticed the two black stripes on her face, laughed silently, without telling her what it was. As if all of that was a cheap gag in a bad comedy, he wanted to draw her attention to this, but did not manage to do so, to talk to her, or, with a clear gesture, bring the message along. The next station was already his, the train stopped and he got off, but not without a last glimpse that should tell her that she should talk to him first or at least stop him. But as expected nothing happened, he just moved on, accompanied with a feeling of sadness that should accompany him for the whole evening. What would have happened - that was his thought-experiment for the following days - if he had taken her hand and drawn dirty things on his face? For a moment, a very short one, he would have given up everything for this unknown person - the existing securities, responsibilities - and exchanged it to the uncertainties of new things. He feels the embarrassing wish to cry, to mourn; even for and especially for everything he had lost so light-headed. But he can't and he tries the dark bread that has been served in the meantime. It tastes sweet, almost like cake.

Until the night falls in, he tries to avoid his acquaintances, to stay away from the main street and walks back to the hotel. Due to the funnel-formed landscape all streets of the village bring him back to the lake. He passes the place where he was in the afternoon and sees her still sitting at the boathouse. He walks towards her, she is not moving, sits in her memorized position that tries to express melancholy and profundity like in a classical painting. She remains in that position like someone that wants to be found like that. Just as he is very close and stands next to her she looks up to him and pats, just like before, the space next to her. He sits down, she lights a cigarette and smokes quietly. Nobody says a word until she extinguishes her cigarette as if she killed a small animal. *A boat-trip?*, she finally asks and points at the small rowboat that is just under the pier. He nods and they put the boat in the water and get on board carefully. There is nothing left from the behavior of hectically acting doctors, running around a corpse that they are trying to reanimate again, although they should know better. In this situation her movements are calm and easy. The unsaid but obvious break-up is happening right now, in this alleged peaceful and quiet moment, while she is constantly looking at the water and the surroundings, getting darker and darker, and he bluntly and silently rowing. With disillusioning disappointment he now sees her in all her humanity and just can't do anything more than finally turn his back on her, to distance himself disappointedly. It might be immature, in this aspect he understands himself even less than others, but he can't do anything about it. He rows further on the lake, passes the middle towards the other bank. Just now he notices the fast growing puddle between their legs, as lake-water slowly enters the inside of the boat. He keeps rowing; the other bank seems closer now than the point from which they have started. Although the water level is rising, slowly but surely, they don't say a word, as if this situation was a strange test of courage. He is tempted to stop, let the boat just sink but keeps rowing. Weaker and weaker he dips the rudder into the water, sometimes on the water, sometimes looking on her. She stares at him as if it was clear what comes next. The bank, what might have been out of their focus because of their limited concentration on each other, is relatively close already; the water level here is much shallower. While she bends forward and puts her right hand on his cheek in a last attempt of tenderness the keel is already aground of the lake and the boat hits the bank, with a dull sound.

Uncle Abdul

ELECTRICAL PLAY
A SURVEY

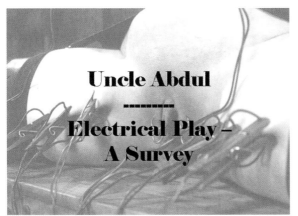

Slide 1

Electrical Play (also known as E-Stim and E-Play) is often associated with BDSM (Bondage and Discipline and Sadism and Masochism) or with solo electrical masturbation, but it isn't well understood by even the users. This survey is an attempt to shed some light and insights into the practice, the theory, and the equipment employed.

My name is Uncle Abdul (see my website at http://www.UncleAbdul.com or e-mail me at UncleAbdul@gMail.com). I've been a licensed, practicing Electrical Engineer with over 50-years of experience in the field. Additionally, I've been a BDSM practitioner for 30 years with many of those spent (especially since 1998) as a lecturer on Electrical Play – presenting in the US and Canada, and read around the world. In 1998 – at the encouragement of publisher friends of mine – I wrote the book, *Juice – Electricity for Pleasure and Pain*. This has been acknowledged throughout the world as the premier primer of Electrical Play safety and practice.

For purposes of this presentation, I'll define:

Slide 2

Here "rewards" can either be pleasurable or unpleasurable. While the pleasurable reward is the goal for many Electrical Play practitioners, the unpleasurable reward is sometimes sought too. This is especially true in the gay Leather community in the pursuit of *machismo*. However, some people find any electrical stimulation unpleasurable.

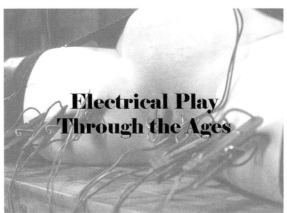

Slide 3

And specifically, for purposes of this presentation, I'm excluding electrical stimulation for the purposes of abuse, torture, and institutional control.

Electrical stimulation of the human body has been around for ages. In ancient times, leg cramps were

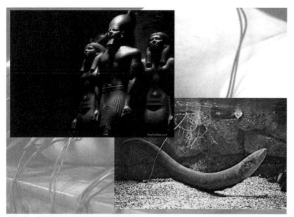

Slide 4

Slide 5

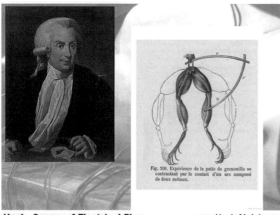

Slide 6

Slide 7

treated with electric fish (Slide 4). In the 18th century, Abbe Jean Antoine Nollet demonstrated current through a series circuit by using a charged Leyden jar to shock 100 of his fellow monks while all were holding hands (Slide 5). Also in the 18th century Luigi Galvani first demonstrated electricity's effects muscle tissue – a frog leg in this case (Slide 6). And in the 20th century, Nikola Tesla, the discover of AC electricity and the inventor of many high voltage machines, often invited friends to experience the sensations of being within a high voltage electric field (Slide 7)

Electricity's Effects on the Body

- **Poorly understood at best**
- **Existing studies done for electrical safety, i.e., injury and death prevention**
- **NO studies done for purposeful stimulation**

Slide 8

No one completely understands the effects of electricity on the human body – and certainly not for purposeful stimulation. And in purposeful stimulation, no studies – medical or scientific – have ever been conducted. This is largely due to fears of litigation and legislation. So the only source of information on purposeful electrical stimulation is in anecdotes amongst its practitioners.

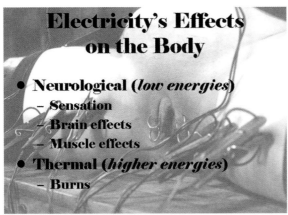

Slide 9

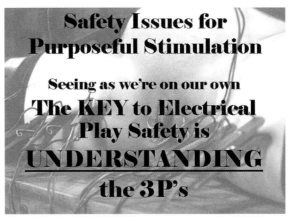

Slide 11

There are some general things that are known as shown in slides 9 and 10.

So, as users, practitioners, and designers of games and other devices utilizing electrical stimulation, what then is need to be understood to use such stimulation safely?

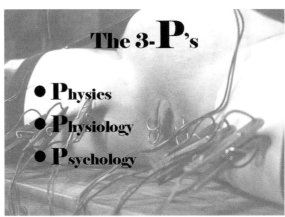

Slide 12

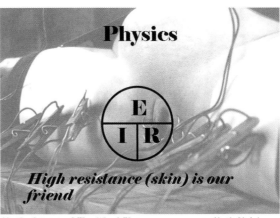

Slide 13

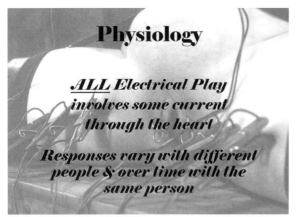

Unc' – Survey of Electrical Play   www.UncleAbdul.com

Slide 14

Unc' – Survey of Electrical Play   www.UncleAbdul.com

Slide 15

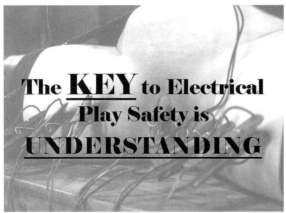

Unc' – Survey of Electrical Play   www.UncleAbdul.com

Slide 16

In my book, *Juice – Electricity for Pleasure and Pain*, I go through the minimum set of knowledge required to use purposeful electrical stimulation safety. Rather than a set of rules to follow, it's more of understanding the 3-P processes involved. Then the participants can assess their own risk(s) of the proposed stimulation method and make an informed decision on the basis of that assessment(s). Here I would refer one to my book to more fully understand these principles. It's really not all that deep or complex. You don't need to be a rocket scientist to do Electrical Play. You just need to understand the 3-P's.

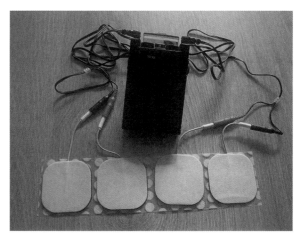

Slide 17

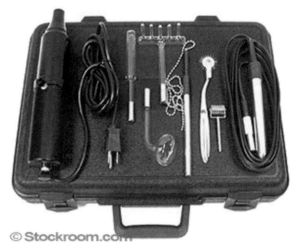

Slide 17a

The above are illustrations of the many, commonly used toys used in Electrical Play. I roughly categorize them into:

- TENS-like Toys – low voltage, low energy toys such as medical TENS Units that generate low voltage waveforms for stimulation (see Slide 17)
- High Voltage Toys – like Violet Wands (see Slide 17a) that generate high voltage, but extremely low energy, fields.
- All-Or-Nothing Toys – like Cattle Prods (see Slide 18) that either produce intense, sharp, brief, but low energy, sensations or nothing at all. There is no middle ground or dimmer controls.

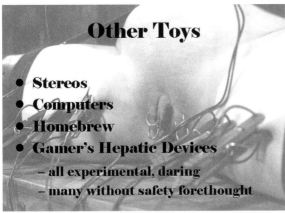

Slide 19

There are other toys that some people use for Electrical Play, but often these are experimental and can be dangerous if considerable thought hasn't been put into their design.

As device designers, what then are the design principles or Mission Statement if you will to build successful devices utilizing electrical stimulation?

Designer's Mission Statement

1. Understand the 3-P's
2. Understand what you're attempting to do
3. Try it on yourself first
4. Don't assume what works for you will work the same for your users
5. **DON'T KILL/HURT the USERS**

Unc' – Survey of Electrical Play   www.UncleAbdul.com

Slide 20

And for those wanting to incorporate automation in their devices, you should especially consider the following:

Automation – Special Considerations

- REMEMBER – The devices generally supply more energy than the body can handle
- *therefore* LIMIT ... CROWBAR ... FAIL SAFE/SOFT
- Check my spec & block diagram at: www.UncleAbdul.com/UAweb113.htm

Unc' – Survey of Electrical Play   www.UncleAbdul.com

Slide 21

Also take a look at Appendix A. *(Page 90)*

Electrical stimulation can successfully be employed in many fun and practical devices. All it takes is understanding the processes and paying attention to good and safe design practices.

COMPUTERS IN THE DUNGEON OR C/BDSM
SOME THOUGHTS ON DESIGN AND CONSTRUCTION / A THINK PIECE

In my previous article I hopefully piqued your interest in C/BDSM. You may have come up with one or two or ??? of your own ideas. What's left? Why implementing them of course. But hold on there! Let's look at what you need to consider to design C/BDSM applications and do them safely. The key word here is safety.

Having done this sort of stuff for a living as a professional electrical engineer, let me share a few rules with you. But first here are some necessary definitions. These definitions are specifically applied to the field of C/BDSM, and they are as follows:

C/BDSM: Computers or devices employed in either E-Stim or E-Play.

Computer: Besides desktop, laptop, and handheld units that we normally associate with computers, I would also include specially designed electronic or mechanical devices. All of them share the following characteristics:

Programmability – i.e., the feature using software, hardware, or hardwiring that allows the computer to perform a specific function or procedure. Generally it also means that the operator can select (change) the different functions or procedures that can be performed.

Inputs – the ability of the computer or device to accept outside signals from electronic or mechanical equipment. This equipment would include operator controls (switches, control knobs, etc.), but it can also be equipment that measures some parameter (like temperature).

Outputs – the ability of the computer or device to deliver signals to outside electronic or mechanical equipment. This equipment would include operator displays (dials, lights, etc.), but it can also be equipment that controls something that moves, delivers a signal that is used by an electrical device, or delivers an electric current or voltage.

Compliance: The ability of an end-effector or a sensor to self adjust to the contour or inadvertent motion of a receiver at the point of application. This is an important safety consideration. (This term is borrowed from the robotics field.)

Device: A single-purpose/function piece of control equipment operating either electrically or mechanically that has the same characteristics as a computer except for programmability.

End Effector: A piece of output equipment specifically used for physically controlling or constraining a receiver or for delivering a physical or electrical sensation to a receiver. Examples would include insertable electrodes, vibrating dildos or butt plugs, heaters, electrodes used to deliver a shock or sensation, an electrical limb stretcher, an automatic cage lock, etc. (This term is also borrowed from the robotics field.)

E-Play: Using direct electrical stimulation on a receiver for purposes of BDSM play.

E-Stim: Using direct electrical stimulation on a receiver for purposes of sexual masturbation.

Fail-Safe: The ability of a computer or device to cease operation of end effectors in a safe and controlled manner in the event of a power failure, operation of a panic switch, or hardware or software failure. The safe and controlled manner of cessation of end effector operation may require that it also go limp.

Fail-Soft: The same definition as in fail-safe except that the computer or device cease its function or program in a predictable and known state so that reset may be more easily facilitated. Again the safe and controlled manner of cessation of end effector operation may require that it also go limp.

Feedback Loop, Closed Loop, or just Loop: A system of sensors and end effectors connected together through computers and/or devices for use in C/BDSM. Such a system will automatically deliver sensations to a receiver through the end effectors in response to signals from the sensors. The operator can make initiation, intervention, and modifications of the functions or procedures in the computer or device. IMPORTANT: See also panic switch.

(to go) Limp: The ability in end effectors and active sensors ceasing operation, due to fail-safe or fail-soft conditions, having its power removed to prevent further operation or stimulation, if preventing further operation or stimulation is the safest condition of that equipment for the receiver. Examples would include removing electrical power from a vibrator or shocking/stimulating electrodes. Also included would be releasing tension in a stretching device or releasing all pressure in a piston. Examples of where removal of power would result in an unsafe condition are the releasing of an electrical latch that causes something or the receiver to drop on and/or injure the receiver.

Operator: The person who directly controls or initiates sensation delivered to a receiver for autoerotic purposes or for BDSM play. Specifically this means through computers, devices, and/or end effectors.
Panic Switch: An input, switch, or other device that will make limp or shut down end effectors under the direct, manual control of an operator in the event a safeword is called or a hazardous, emergency, or unforeseen condition exists that would result in injury (physical, psychological, or spiritual) to a receiver. The panic switch would override any controls done by the computer or device.

Receiver: The person who receives through computers, devices, and/or end effectors the sensations controlled by the operator for autoerotic purposes or BDSM play.

Reset: The operation and the ability of a computer or device to go from its stopped fail-safe condition or from its predictable, known fail-soft state to the beginning of its program or the start of its function on specific initiation of the operator. During the transition from this condition or state, the end effectors and active sensors will remain or be brought into a condition that is safe for both the receiver and the operator. This may even require that the operator remove the end effector or the active sensor from the receiver before the operator resets the computer or device.

Safe: Specifically a state where both a receiver and the operator is not at risk or subjected to physical, emotional, psychological, or spiritual injury to their persons and/or bodies beyond those limits sanely and non-exploitatively negotiated beforehand and which do not exceed known and published medical limits. Unsafe is of course the opposite of safe.

Sensor: A piece of input equipment that is used to measure some parameter such as temperature, blood pressure, breathing rate, skin conductivity, etc. It may also be an input measurement device. Sensors may be active, i.e., require external power or pressure to operate, or passive, i.e., not require external power or pressure to operate.

With these definitions under our belt, it is now time to consider what one needs to have in mind in designing C/BDSM equipment and systems. Again the following rules are based on my experience in designing systems for industry. And again they are specifically adapted to C/BDSM equipment. Here goes:

1. All such C/BDSM equipment, systems, sensors, end effectors, etc. – whether they be especially designed and/or invented for the purpose or adapted equipment – should always be designed or applied from the standpoint of safety for both the receiver and the operator. You'll note from the C/BDSM definition of safe that it produce no more injury than negotiated for. It also says that it should produce no injury that exceeds medical guidelines. For a C/BDSM system built to be used on a variety of receivers, this might mean that it be capable of producing no more than a first degree burn over no more than 1-per cent of the receiver's body. Certainly

in an electrical shocking device you wouldn't want to go beyond the Underwriters Laboratories™ current limits for causing muscle spasms. (Generally this is about 10 milliamps.) But if, say, you had an automatic caning machine, and you negotiated using it with a receiver, you shouldn't if the receiver were a hemophiliac. Similarly you wouldn't want to use an electrical toy with a receiver with a known heart condition or who had been a victim of electrical torture.

It is always the responsibility of both the operator and the receiver to assure themselves of the safety of the C/BDSM equipment or system that they're using – regardless of whether it was self-manufactured or manufactured by others. It is also their responsibility to see that said equipment or system is operated in accordance with the manufacturer's instructions, and that the limits of the equipment or system not be exceeded. As an example, you may have tested your violet wand beforehand, but for God's sake don't play with it in the shower.

2. As many of us know from painful experience, computers and devices have a habit of 'freezing up' – i.e., the program gets stuck at one point and refuses to work any further. Your only recourse is CTRL+ALT+DEL or recycling the power switch. Think about this when you design your C/BDSM equipment. Much of an engineer's work is thinking about the 'what if' issues in designing something. You should too.

Here I specifically direct you to the definitions for fail-safe, fail-soft, and reset. Generally engineers try to design their systems so that if their computer or device fails or halts in its program, the system is left at a safe condition. And when it is time to continue the operation after the operator had assured themself that the failure has been fixed and that it's safe to restart the system, then a specific reset operation – like pushing a RESET button – is required by an operator. And when the system starts operating again,

engineers make damn sure that it starts safely – without damage to personnel or equipment.

Here it is important to know and understand what the risks and potential damages are (the 'what if' questions). In C/BDSM design think about what your end effector will do if the power fails or the computer hangs up. Also consider how the system should restart when the failure is fixed and unsafe conditions are cleared.

3. It should go without saying, ALWAYS PROPERLY REPAIR FAILURES THAT OCCUR IN C/BDSM EQUIPMENT before restarting it or using it again. Similarly, ALWAYS MAKE SURE THAT THE OPERATOR AND THE RECEIVER ARE SAFE before restarting C/BDSM play.

4. C/BDSM systems and equipment should always have some sort of panic switch under the direct control and within reaching distance of the operator. Other panic switches for the receiver or a dungeon monitor can be designed into the C/BDSM equipment, but these other switches must operate in the same way and through the same means as the operator's panic switch. And while an operator or receiver will not be able to react fast enough to prevent injury by excessive current in direct electrical stimulation, a panic switch can prevent further injury.

5. Panic switches and other emergency shutdown devices should directly cause the removal of power or pressure from end effectors and active sensors (see para. 7). Such power removal operation should not work through the computer or device nor depend on the operation of the computer or device. It should be independent of the operation of the computer or device.

The panic switch(es) and emergency shutdown device may report its state to the computer or device in their design. The design may incorporate this input signal to allow the computer or device to achieve a fail-soft state.

6. In C/BDSM systems and equipment that are programmed to perform a sequence of events or to control a feedback loop of sensors and end effectors, the system or equipment should incorporate a shutdown timer wherein:

    a) It must receive a positive, periodic signal from the computer or device to continue to operate

    b) The shutdown timer will operate as long as it receives this signal and it is energized

    c) Loss of the periodic signal from the computer or device or it's energizing power for longer than a safe period of time* shall cause an emergency shutdown of the C/BDSM system or equipment (see para. 7).

    d) Re-energization of the shutdown timer(s) shall only occur during manual reset

    * This safe period is the minimum cycle or event time required by the C/BDSM system or equipment. It is highly recommended that it not exceed 16 milliseconds.

    The shutdown timer may report it's state to the computer or device in their design, but it shall not cause the operation of an emergency shutdown through the computer or device. The design may incorporate this input signal to allow the computer or device to achieve a fail-soft state.

7. End effectors and active sensors should be designed in a way that it requires the presence of power (electrical or pressure) to make it operate as designed. Also they should be designed so that removal of that power or pressure shall:

    a) cause its operation to cease

    b) assume a safe state, position, or condition

    c) remain at this safe state, position, or condition until the equipment is specifically reset by the operator

    d) not go into an unsafe state, position, or condition if the power or pressure is reapplied upon reset

Additionally, it is highly desirable that all end effectors go limp upon removal of power or pressure. In pressure systems, this may require that the equipment be vented to the atmosphere or to a return reservoir.

8. It is recommended that upon reset and restart of a C/BDSM system that the system and equipment operate at their lowest design settings and operating points. This is intended to allow the receiver to physically and mentally prepare for and to accommodate to the sensations imparted.

9. Where practicable, C/BDSM systems and equipment shall be designed in a way that a failure of any single component of the system or equipment shall not lead to an unsafe condition for the operator or receiver.

10. C/BDSM systems and equipment shall be designed and built in a manner that neither the operator nor the receiver be exposed to hazardous voltages or currents or to moving components within the control and operating portions of the system and equipment that would cause serious injury or death to the operator or receiver or to bystanders not involved in the C/BDSM play. Here a designer should reference the National Electrical Code (NEC), the Underwriters Laboratories™ standards and specifications, and other applicable documents.

11. End effectors and sensors shall be designed, specified, and constructed with compliance (see definition) in mind. Cases of emergency or inadvertent moves must be taken into consideration. As an example, consider the situation where a vibrating dildo affixed to a rod anchored to the floor is inserted into a receiver suspended above it. What happens if the suspension fails?

You don't want the receiver to be injured by impalement.

12. Mechanical components incorporated in C/BDSM equipment should be designed and/or specified with a minimum safety factor either of 3 or of the minimum value specified for that component, whichever is greater.

13. End effectors and sensors shall not have any sharp edges, burrs, or deformities that can cause injury to the operator or the receiver. This is especially critical for end effectors and sensors that are intended for insertion into a body.

14. End effectors and sensors – especially those that are intended for insertion into a body – should be constructed of such material that is neither injurious nor poisonous to human beings.

15. End effectors and sensors that are intended for insertion into a body and utilizes an electrical cord or pressure tubing for withdrawal should be designed and constructed in such a manner that neither the tube or wire does not break or separate nor the equipment break or disassemble during the withdrawal. Also such end effectors and sensors should be sufficiently sealed to prevent bodily fluids from their interiors. And their design and construction shall be such that cavities or voids on their contact surfaces shall be minimized. (Remember, muscle tension during times of stress or excitement in a receiver may at times be more than normal.)

16. It is recommended that end effectors and sensors be designed and constructed in a manner to allow for cleaning in a dishwasher with 140° Fahrenheit (60° Centigrade) water.

17. End effectors and sensors shall be designed and constructed to meet or exceed the requirements of the environmental conditions to which they'll be subjected.

18. C/BDSM systems and equipment shall be designed and built in a manner that the temperatures of any of its exposed surfaces shall not exceed 100° Fahrenheit (37.8° Centigrade).

I presented these as a series of thoughts and considerations based on my experience. They in no way can be considered definitive not official. The designers and users must assess their own risks and accept full responsibility for their use of C/BDSM. And users – both operators and receivers – should look at these issues even if they use systems and equipment supplied by a manufacturer.

I would also like to hear from users and experimenters that either are contemplating or actually playing with C/BDSM. You may contact me at UncleAbdul@gMail.com or write me at:

Uncle Abdul
530 Showers Drive
Building 7, #224
Mountain View, California 94040

Chao-4-Now
Unc'

Elle Mehrmand (Echolalia Azalee) and Micha Cárdenas (Azdel Slade)

EROTIC ELECTROSYMBIOTIC ENCOUNTERS

I remember walking in the street in the rain, with echolalia under an umbrella, hunting for a medical supply store, since I still have all the memories from my fleshly body. We need to have an erotic encounter, but we arrived in Bogota without our gear. We stop at a pharmacy storefront that opens on the street. "Donde podemos buscar una tienda de equipos medicos?" I ask. "En la calle septima, tres cuadras mas," the pharmacist tells us in her white lab coat, endless rows of pharmaceuticals behind her like Borges' library of drugs. First we have to find a stethoscope, and then a piezoelectric sensor, some wire, a soldering iron and a 1/8" jack.

Amazingly, we find this medical supply store in a byzantine urban metropolis we aren't familiar with. The city directs us to it, like an eddy, it lies in a hallway off the main street at the bottom of a towering office building, past electronics stores and restaurants selling sausages and arepas. Once we arrive, we ask about stethoscopes, we want the cheapest one. The woman at the counter seems slightly confused, slightly suspicious. I pay with a magnetized plastic strip that has my legal name on it, Micha. We leave with two stethoscopes and a lot of medical tape.

Azdel and I press on through the electronics district, where all the electronics stores are stacked and crammed into a few city blocks, the bright LED's and signs hallucinate my vision with every step. The store fronts have too many items in the window to recognize any one, millions of wires, adapters, speakers, lights and scrolling LED displays. After too many stores, and too many epileptic lights, we settle on a fifty cent piezoelectric sensor, to be our contact mic.

Back in our hotel we tightly tape the sensor to the stethoscope, a makeshift version of the gear we've been working on at home that's sitting on our desk. The sensor is bound and mummified with medical tape and is attached to the stethoscope, we plug-in and start kissing. This was the first time we heard each others live heart beats, new intimate encounters arise.

After getting back from Bogota, we're sending our meat bodies to San Francisco on another plane. We need to engage in another erotic moment of kissing, touching and rubbing, so we start looking for a good public space. On the grid again, I go to the Aftermath club in Insilico. It's daytime now, so it's deserted, although one could switch the environment/ time setting with a click. The DANCE DANCE sign scrolls by, the images of x-rays of skulls and other body parts flash rapidly on the screen above the DJ booth. The screen on the floor flickers, lines of color look like raw data, no signal to control the pulsing. It is blue, white, grey, and hard to distinguish. The silent arrays of speakers hang from chains, but tonight they'll be alive with the sounds of our bodies. This club is perfect; the light spins and reflects quickly off of my mechahooves. I can't wait to see echolalia's pixelated skin under these lights. I walk to the Insilico Transit System screen, a GEMINI Cybernetics Teleporter station, touch a destination and I'm gone.

Micha and Elle wake us up, they're at the center for sex and culture, we were laying down in our lesbian love bed in the skybox waiting for them to login. I pan over and see Azdel standing next to me, her mechahooves are glowing, and her still bloody bandaged tail sways in excitement for our erotic encounter. My ears won't stop twitching; sometimes I get anxious when we take our clothes off in front of strangers. The projection is ready, and we have entered our full scale body projection, the only time we are proportional to our fleshy selves. Elle starts to take my clothes off, leaving my mechanical leg and transparent cyber bra on. I wish she would put on my strap on so I could fuck Azdel, but I guess that's for another performance. Micha takes off Azdel's bra and tiny skirt, and she gives her the long braided glowing hair, I love how her hair shimmers when we kiss.

We stand face to face while Micha and Elle get the final details ready. In front of the murmuring audience, they open their laptops and run all the necessary software to create this erotic moment, the Puredata software to read the live biometric data, the Universal Mixer to send the sound into the virtual audio device and the Second Life client. Micha and Elle pour a bit of water into their hands, each in turn wetting their sensors to get the proper conductivity and contact. Elle takes off her dress and straps on the soft electrocardiogram transmitter. The USB cable slips into its hole and the lights on the electronics come alive, flashing with anticipation. They hover over their laptops configuring the last details of the heart rate patch, the traced line of the heart beat graph looks as it should, the data is pouring in over the serial interface, and once the sample chunk size is set, the sound of their heart beats starts pounding out of the speakers and over the audience, out of our bodies and into the space of Second Life. The electrical impulses in their hearts travel through the intimate conducting system to the heart muscle, stimulating the myocardial muscle fibres to contract inducing systole.[1] The biometric sensors detect these impulses and they are translated into sound and motion in our networked reality of Second Life. With all the links in place, our realities are mixed, we begin kissing and I feel echolalia's spinning mechanical leg rub up against me. Her cat ears feel soft on my hands as our digital tongues telematically connect.

1   ECG- simplified. Aswini Kumar M.D, http://www.drsiordia.com/en/electrocardiography/

We are technésexual. Our sexualities do not fit into any reductive formulations of Lesbian/Gay/Bisexual/Transgender. Those old categories were useful for building coalitions and fighting back, but they don't map well onto our neko bodies, our hybrid cat-human assemblages of off the shelf biomods, mechatronic limbs and body hacks, or our mixing of physical/virtual flesh, either. My tail doesn't wag for any "same sex" partner, but for the electricity of my love's body. Our sex toys are not just dildos, but also 9V batteries, sticky electrodes, hand soldered components and DIY biometric sensors. And this is no fetish, any more than heterosexuality is a fetish, no disorder any more than cisgender is a disorder. The quick flashing of the LED's on my partner's microcontroller with custom coded firmware is to me like a quickening of the breath. Even more, the sound of her heartbeat, channeled directly into my ear, loud as any pulsating soundtrack, is what excites me most. I know just how to rub her clit to raise her heart rate or kiss her deeply to lower it. And I know that when she grabs my feminine cock that my heart rate changes, for her to know and anyone else in ear's range, across two realities, physical and virtual.

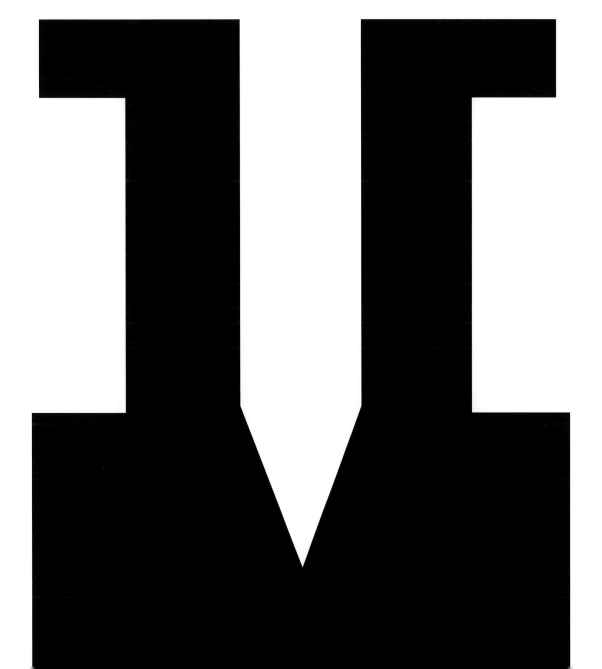

Douglas Bryan LeConte-Spink

SCIENCE FICTION & CROSS-SPECIES SEXUALITY
THE UN-CROSSED BOUNDARY

Johannes Grenzfurthner: A big applause for Douglas Spink, our next speaker. Douglas will talk about "Science Fiction & Cross-Species Sexuality: The Un-Crossed Boundary," and I hope you will enjoy the talk. I do hope we have some controversy later on...

Douglas Spink: Oh, I'm good at that... I'm the controversy guy, so... [mic noise]. Better? Hello. Test, test, test. It seems good. All right.... I can hear it squealing just...

Johannes: Oh, no. it's [microphone: squeeeeeeek]. Ouch!

Douglas: Get closer.

Johannes: The farther you are...

Douglas: The more it squeals? OK. Well, there, that just shows how much I know.

Anyway, I'm one of the - I hate to say 'anti-singularity' people, because I'm not anti-singularity - but I'll out myself as one of the people who posts pages and pages of responses to lots of the articles on H+ (hplusmagazine.com), which I think is a great magazine; I want to say that right off the bat. I'm the guy who seems to be the... well, yeah... my position actually perhaps more nuanced and complex than what my responses to all of the articles on bio-computation (and other stuff like that) might otherwise suggest. So, what is my position on those questions?

First, a bit of history: my official academic foundation is in complex systems theory in the past, although I've been "almost finishing" my doctorate in the subject for nearly seven years now. I'm a little bit lapsed it's safe to say. Yeah... officially lapsed at this point, I believe is a more accurate description. In the doctoral work I have completed, I have focused primarily on algorithmic information theory, network complexity, and set-theoretic reconstructability analysis. The idea is that this foundational work will lead to my dissertation topic, which is "Quantitative Theories of Consciousness" - though I've still a ways to go before I'll be ready to bite of that particular topic and do it full justice (in recent years, I've veered more into cross-species epistemology but that's a different talk for a different conference, perhaps).

Given my formal academic background in systems theory, one of the critiques that I have of some of the simpler versions of singularity perspectives is that they tend to ignore the fact that we've basically hit a complexity wall in the world of software engineering. No matter how much faster our computers are getting, we don't really know how to make more complex code structures that actually _work_. We all know this is true; it's empirically true. We even know why it's true. For those who have read Wolfram's "A New Kind of Science," for example, it all ontologically fits. We just don't currently have clever ways to resolve these issues; in fact, we don't even have proven theoretical frameworks to undergrid what might be promising _future_ approaches. It's intractable, absent some serious conceptual breakthroughs that don't exist yet. Indeed the problem gets far, far worse when we consider quantum computing - for which we have no algorithmic models currently. Hardware heaven, software hell.

Kurzweil - bless his heart, he's a very interesting guy. No question, he's a brilliant guy... but the singularity perspective just seems to me like it's a little bit built on a house of cards, in terms of the empirical, quantitative data upon which it rests. Time will tell...

Anyway, I was one of those people who have been annoyingly responsive to lots of those articles. I didn't want that fact to be lurking out there, as if I were hiding behind a screen name whilst knowing that I've stirred drama in comment postings there. That said, H+ Magazine is a wonderful resource - in part because it's asking strong questions; that's obviously a crucial role to play in any genuinely path breaking project.

That clarified, I'd now like to turn to the official subject of my talk today. In this talk, I'm using the subject of cross-species sexuality as presented in the science fiction literature for a somewhat similar set of reasons. Specifically, I'm looking to ask good questions - and to generate interesting perspectives: no more, no less.

I should say right off the bat that there's some spoilers in this, so if you've not read some of the more abstract hard science fiction literature and you're completely panicked that you might find out some plot point of a book published in the 1950s, and - oh my God! - it's going to ruin your life and cause you to going to go bonkers with anger... well, then you should perhaps leave now, because I might drop spoilers. I could give you a list of the books - if you really need to know that in advance - but it's really not that bad. Still: spoiler alert.

Tactically, the notes I'm using for this talk have been posted with the Twitter hashtag earlier today, so if anybody wants to have a full copy of the notes that I'm using to do the talk, it's already there; you don't have to take notes, or anything like that (unless you want to). Also, in this talk, there's no presentation stuff: neither fancy slides nor PowerPoint presentations. Instead, it's simply me talking through some subjects, working from those notes. This is one process I know how to follow without making a hash of things, so I'll stick with that. Here goes...

As a species, we use stories for more than just entertainment. I think that's likely not a surprise to anybody here, is it? In fact, it turns out that we use stories as a critical component of our cognitive framework: to allow us to make sense of the world in which we live and - in particular - the social world in which we live. We use stories as efficient tools for the continuous dissemination of inherited wisdom from generation to generation. We use stories, basically, as a form of compressed algorithmic information within our species, because it's a lot more efficient to learn an algorithm than it is to try to memorize by brute force all the possible outcomes of all the possible things. If

you think about Aesop's Fables, with the Fox and the Hare and so on and so forth, they teach us important lessons by encapsulating important concepts within a highly compressed form. Algorithms.

Science fiction, I argue, does this for our culture nowadays - it does that quite a bit. It is a tool for the exploration of new ideas... but it's also a normative tool for teaching us important things about what it means to be a technological species. It does that in a way that is a lot more powerful than I think we tend to give it credit, because it has become so much a part of our cultural framework. The trees are lost within the forest, as it were.

Other species do use stories, of course, to teach each other. But, in those cases it tends to be more in the form of first person than third person. I was thinking on the plane, on the way down this morning, of examples of the dogs in my family teaching each other tricks - one to another... usually bad things, or at least things I don't want them to do. If one of them figures out a new way to do something that's particularly clever then - before you know it - the

trick just spreads like wildfire, because they watch each other and they learn through observation. In short, because they are a social species - just as humans are a social species.

We're mammals, too, and we also learn by watching each other. Stories. But we can do that at a second order. We can learn by listening to each other's stories - not just watching "physical stories." Further, we can start making up stories that didn't actually happen, so that we can teach people stories and tricks about stuff that never 'actually' took place.

This is the capacity from which human culture emerges; it is the core of human culture. That's why reading science fiction and looking for interesting perspectives, I think, is fully worthwhile. Because these sci-fi stories are our modern stories and they have been for a number of decades - especially in the technological realm.

For that reason, I propose that really what I'm doing today is presenting a Monte Carlo simulation using the science fiction literature at a tool for exploring the landscape of probability - or potentiality - within the question of cross-species sexuality in our culture today. That's my formal statement of theoretic intent. In reality, I'm just going to talk about sci fi and different approaches to cross-species questions and what that all about sci-fi and sexuality - and sexuality, in general, in our culture.

I'm sticking specifically to science fiction here, because if I started to bring 'fantasy' literature in, well... the topic just gets too big. It gets too diffuse, as well, and doesn't have the sharp edge that science fiction has. I think fantasy is more about our history: in a nutshell, I think fantasy talks about who we are via different prisms through which we can look. Then, science fiction - super over-generalized - talks about who we think we are going to be in the future; these are both ways of talking about who we are today, of course, but from different foundational assumptions.

But when I look through the cross species fantasy literature, it's just absolutely overflowing. I mean really, truly overflowing. I couldn't even start to do an hour presentation on it to do it justice. I could do an hour presentation on Robin Hobb's "Assassin's Apprentice" trilogy... and not get done with that. Whereas, in the science fiction literature - and in decades of reading and taking notes on these topics - I have three pages of raw data. So, this is a targeted approach. Further, I think that's what makes it quite interesting as much for what we _don't_ find as for what we do.

What I'm presenting today is just science fiction. That means I've left out most of Piers Anthony's work, which has got a lot of interesting stuff in terms of cross-species considerations. No doubt, he's an interesting writer - and he writes on that sci-fi/fantasy border. A lot of people do, and so it's hard to really say... For example, David Brin: I think he's more science fiction in most of his stuff. To be honest, I've left his stuff out because I gave his books away to a

friend of mine - and I couldn't get them back! I didn't mean to leave Brin out, because he is an important component. He's got a lot of dolphin stuff that is actually pretty cutting-edge in how he looks at things. So there you go: reality intrudes...

That's most of my introduction. Oh, and regards to more personal questions and personal perspectives: so that I can get through the content here, I'd like to hold those to the site. That said, my unique perspective on this presentation, perhaps, is that I'm a little bit equivalent to Sack's "Anthropologist on Mars" - a stranger in a somewhat strange land. That's because most of my bonds, throughout my life since I was a kid, have been undeniably cross-species.

I grew up with horses, riding horses, competing on horses in a 'horse-y family.' I've personally found on the mild end of the autistic spectrum - Asperger's Syndrome - and I've always bonded more naturally with the four-leggers than with two leggers, throughout my whole life. So, I do sit astride the species boundary - more than a little bit. I'm definitely a human and do human stuff - it's not a question of feeling misplaced in the species, for example.

In sum, I tend to come at questions of trans-species epistemology from a slightly different - often orthogonal - angle. I spend all day - every day - with non-humans... far more time than with humans, very much by choice. There are weeks at a time that I don't actually seen human beings. I'm certainly not alone in this; best research data are that about 1% of the human population has this gift, this ability to genuinely bond empathetically beyond the limits of our own species. It's something I've come to call "Deep Symbiosis" - which I think is useful date to understand the frame of reference from which I approach the topic of this presentation.

Indeed, from that perspective, watching human beings as we talk about sex is always fascinating to me: how confused the whole conversation can be, how strongly impacted by situatedness; humans are really not self-aware about their sexuality... not at all. Blinders on.

And this self-blindness is truly difficult to place in context of the vast swath of life on our planet. Let us stop and think about this: the majority of species (in terms of number, though maybe not biomass) on our planet reproduce sexually. Not just the animals - which is sort of interesting. After all, the vast majority of plants do, as well. So, sex is not unique to us as a species. Sex is not unique to animals. Indeed, sex is not even unique to animals. The majority of plants in the plant kingdom - other than like algae and such which are truly ancient - also reproduce sexually.

In sum, sex is a clever trick that life figured out a long time ago - and it has since spun out into countless different variations. From a structural functional perspective, it's essential to natural selection. In fact, sexual selection was explicitly recognized - in Darwin's initial presentation of his research results - as an equal force to natural selection: equal but separate. We've somewhat lost track of that, in more recent theoretical and mathematical models of evolutionary process, in that we now tend to think of natural selection - "survival of the fittest" - as the basal driver of evolution. In that (mistaken) view, sex is just a reproductive procedure, shorn of evolutionary centrality. However, that misunderstanding is not empirically correct... and it's not even in line with Darwin's own perspective. The fact is, sex is front and centre to how multi-cellular life works on this planet.

[background noise] Sorry, I keep thinking they're coming for me... I'm always looking over my shoulder {an ironic comment, given the flow of events since this presentation was made!}.

Given that factual foundation, we can accurately observe that humans fit squarely into this larger constellation of sexuality and sexual behavior. Even if we leave the plants out for now (and just refer to the animal kingdom), there's very little that humans do - sexually - that is unique in any meaningful, substantive, qualitative way. Perhaps 20 years ago such a

statement could have been preemptively tagged as 'controversial,' but at this point I'm making a statement of fact based on wide, cross-disciplinary scientific consensus. Our sexuality is part of the warp and weave of animal sexuality in general.

Well, perhaps I'm overoptimistic about such consensus. I suppose that one can make this cross-species sexual observation - viz., that we're part of the larger world of animal sexuality, not separate and distinct from it - and a recalcitrant rump of people will still persist in giving embittered argument: the zealots, the religious fundamentalists, the angry demagogues... in short, those who are simply scared by the real world and thus unable to see themselves as part of the living world - rather than forever, ontologically isolated from it. Those people are still out there; they're still loud, still dangerous, and still looking for targets to use in hysterical, witch hunt-style attacks against those who speak truth. It's best to remember that, as ignorance of the power of the bigoted backlash can be a dangerous mistake. In the past, they burned heretics at the stake for making such statements; nowadays it's hate-mob hysterics and prosecutorial use of hijacked legal systems that serves as their torches and pitchforks... it's best to remember that.

But, at the least we can still observe that it's not as bad as it was in the past - I certainly hope it's not. I remember when I was a kid, having veterinarians argue with me that animals "don't enjoy sex" (as a categorical statement, "animals" versus "humans" is always delightfully counter-logical and illustrative at the same time... after all, are humans thus plants if not animals?). It was a fact - or so they said: they just don't like it, not one whit. It's been proved, so they claimed: they don't enjoy sex, period. They only "do it" (such a quaint, teenage term) because they have to. It's just an urge, just a pre-wired, mindless, simple behaviorist thing. Skinner would be proud.

Nowadays, it's just funny that you could actually say such nonsense with straight face, isn't it? How could anyone - anyone - who has ever spent meaningful time with other mammalian species say something so utterly vacuous? Don't they watch the Discovery Channel? Or, is that censored nowadays to elide any reference to the pleasurable facts of sex? I don't know, I haven't had a TV for more than twenty years. I do know that, even today, you'll get the anti-factual nonsense from some actual veterinarians - which never ceases to amaze me! They'll tell you these things, these "facts" that aren't true - and sit back, expecting their non-facts to be accepted as valid. No citations, no objective support, nothing - just those quasi-religious statements of inanity. A frightening thing to see... made all the more so when we consider the terrible consequences such intentional blindness can - and does - cause. Simply put, the ignorance that some humans are still able to conjure forth - facts be damned - when it comes to non-human sexuality is a mirror reflection of our profound ignorance of ourselves and of our obviously intertwined place in the living world.

When humans get to sexuality...well, it's a little bit like we're trying to look at ourselves in a warped mirror... a form of trying to see the outside our closed universe but hopelessly using only the cyclical lens of a recursive loop. It's hard for us to see - to truly see - and understand our own sexuality as it exists in the spectrum of the wider living world; it's hard because it's so much a part of who we are (same goes for most all the social mammals, but we'll get to that later). For this - and other - reasons, we have a terribly difficult time being objective about our sexuality, let alone about sexuality as a more general question of study when it transcends the ersatz species barrier.

Consequently - and now stepping back for a broader view - as we try to look at sexuality through the prism of how other species integrate it into their social, emotional, spiritual, and cognitive selves, the power and importance of this analytic stance becomes clear. When we do so, we see that how we deal with the question of sexuality as it transcends species boundaries illuminates our most fundamental assumptions about other species... not to mention about how they deal with the question from their own frames of reference. So, when it comes to questions of sexuality,

I think it's useful to take this broader perspective because when we loop back to look at ourselves, it enables us to break out of our recursive trap. We're no longer staring at our own enchanting reflections in the mirror, no longer steadfastly focused solely on our own species' navel. In short, it puts us into the landscape of a more grounded framework.

We take this cross-species stance rather than just floating around out there in the vacuity of the "sex is this big imponderable thing that's uniquely human and uniquely complicated and uniquely frustrating and uniquely... etc." It's not unique. Rather, it's a frustrating, challenging, complicating, central part of who we are. But no: it is not specific to our species at all, not unique in that regard.

Most of the assumptions that I think the average human person (i.e. someone who isn't following the ethological literature on other species' reproductive processes, and so on) carries around have to do with thinking that there's a genuine uniqueness in how humans express themselves sexually, with respect to other species. This shows up in the sci-fi literature, as well, and it's of the themes that I shall continue to touch on as we go forward with our explorations today.

So... what do cross-species considerations have to do with the parallel - but distinct - questions of the intersection of sex & technology? I mean that this is, after all, a sex & technology conference. And, to get that that question of relevance, I'm going to read a quote from something from which we're going to follow up from there. Once that's done, we'll jump into some selected quotes from the sci-fi books that I'm going to use as examples in support of my deeper thesis.

To start, I'll quote from a work titled "The Companion Species Manifesto." It's by Donna Haraway, who is also the author of "The Cyborg Manifesto" - a work that was quite powerful when it was published in 1985. That earlier work was a kind of statement of inculcated feminist, radicalist, cyborg politics... I guess you could say. The earlier is also interesting book and a powerful book - even today - to encounter. But, nowadays, Haraway has moved past cyborg conceptualizations and into companion species considerations, because she's decided it's the implicit, inevitable next step. That is, considerations of the cyborg now must blur into those of the companion species. In doing so, she goes back to explain some of her early work:

"Cyborgs are cybernetic organisms named in 1960 in the context of the space race, the Cold War, and the imperialist fantasies of techno-humanism built into policy and research projects. When writing "The Cyborg Manifesto," I tried to inhabit cyborgs critically, i.e., neither in celebration nor condemnation, but in the spirit of ironic appropriation for ends never envisioned by the space warriors themselves."

- and -

"Telling a story of co habitation, co evolution, and embodied cross species sociality, this present manifesto asks which of two cobbled together figures - cyborgs or companion species - might more fruitfully inform livable politics of ontology in current life worlds. These figures are hardly polar opposites. Cyborgs and companion species each bring together the human and non-human, the organic and the technological, carbon and silicon, freedom and structure, history and myth, the rich and the poor, the state and the subject, diversity and depletion, modernity and post-modernity, and nature and culture in unexpected ways. Besides, neither a cyborg nor a companion animal pleases the pure of heart who long for better protected species boundaries and sterilization of category deviants."

That's Haraway; she's essentially taking the same position that I'm seeking to stake out today: that cross-species questions and cyborg questions are blurring into each other, inescapably. In fact, the sci-fi literature quite directly speaks to this; it has itself pursued these questions deep into the possibility landscape of new ideas.

Next, what I'm going to do is read some quotes from a broad spectrum of science fiction work, mostly towards the 'harder' edge of the spectrum. I've set up a 2x2 matrix: it's so much fun to do 2x2 matrices, eh? As humans, we so much tend to think in "base 2" terms (Levi-Strauss would concur). Binary.

If I had a whiteboard behind me, I'd draw a vertical line. This would be my "politics" line. The bottom of the vertical line would be marked conservative: basically old-school, old-fashioned historical politics. Then, at the top of the vertical line, we've more of an open-ended, open-minded, socially unconstrained view of social relations. Flexible, as opposed to conservative.

Next, for our horizontal line we have, on one side: individualistic, which is essentially an inward-focused conception of what sexuality is about. And then, at the other end to that line is the "Other" - as in a capital 'O'. Taken together, this 2x2 matrix maps out, I think, our conceptual landscape. With it, we can map various sci-fi authors when they're talking about sexuality in general, but cross-species sexuality in particular.

Ok, let's jump into the quotes... starting with one from Greg Egan, from a short story called "Wang's Carpets" (written in 1995) - he makes a good starting point for what we're doing today, with this excerpt:

"The real test wouldn't come until a diaspora or the Gleisner robots finally encountered conscious aliens... minds entirely unrelated to humanity, observing and explaining the universe that human thought had supposedly built. Most ACs [I'll explain the term below] had come right out and declared such a find impossible; it was the sole falsifiable prediction of their hypothesis. Alien consciousness - as opposed to mere alien life - would always build itself a separate universe because the chance of two unrelated forms of self-awareness concocting exactly the same physics and the same cosmology was infinitesimal, and any alien biosphere that seemed capable of evolving consciousness would simply never do so."

Here, "AC" is a term he's coined: anthrocosmologist. It implies a version of imagining that human beings - by being conscious - create the universe around themselves... reality as an artifact of consciousness, a different flavor of the "there's a computer program running, and we are all in the program; the Matrix" ontology.

Anthrocosmology is, in actuality, a real component of some deeper fringes of mathematics. It's present in some now-discredited interpretations of quantum mechanics: that by observing the universe we create universe. If you take that to extremes, the universe is sort of an artifact of human consciousness... which kinds of implies that only humans that are conscious, because otherwise you'd have a bunch of different universes. So... what if we bump into other conscious species? Do we all live in different universes? Do we blow apart into separate realities?

The interesting question - and the reason that I bring Egan in, is that he's brought forth this clever questioning of a deep cosmological issue. To wit: there's - of course - conscious life all around us every day, on our planet... consciousness that isn't human. Ten years ago, that was controversial: "wait... humans are the only conscious species!" No, they're not - and that's now been shown true by cognitive ethology, quite definitively. The "sphere of consciousness" is getting bigger and bigger... depending on how you define it, who's doing the research, et cetera. But, this idea of consciousness being uniquely human has fallen away. We live on a planet that has many conscious species... perhaps conscious in different ways and to various degrees than us... all mammalian species, at a minimum.

So, to imagine that we have to go to other worlds to find other consciousness - to see how they'd view the universe... it's ironic, since we live in a world with other consciousnesses whom we just ignore (or kill or whatever) on a regular basis. We've put these blinkers on, and said we're the only thing that matters on this planet... that we must needs go elsewhere to find the Other; it's very odd, eh? I think Egan himself skips

over this in his 1995 work, though you'll see some temporal progression in his perspective, when we see him trying to get his head around the classic question of what happens when you meet aliens.

My rejoinder to the fallacy is simple: we live with aliens already here on our own planet - viz., the other conscious species - and we don't know what to do with them! Heaven knows what's going to happen if we meet "real" aliens out there; we probably wouldn't even recognize them as alive. Worse yet, with us already living on a planet with conscious, complex sentient "Others" what we do is we kill them, we do horrific things to them. If we meet other species from other worlds, we're going to have a damn hard time explaining exactly how we've behaved here.

Ok, back to more quotes. I'll tell you the author and the year and then read the quote, and we'll see if anybody can place that quote in my 2x2 matrix... the one "behind me" on the virtual whiteboard. Again, the vertical line is social politics; strict to open-ended. The horizontal line is self-focused sexual ontology versus "Other" sexual ontology. That horizontal line points out the obvious dichotomous split - and, actually the singularity discussion hit on this - that the ultimate form of sex is either with another version of yourself... or it's having sex with the ultimate Other. Well, it's probably not purely an either/or, but rather two different _ways_ of looking at sexual interaction: is sexual interaction about that bond with the Other, or is sexual interaction about improving the quality of the orgasm for the Self? You could almost say that was a male/female question, perhaps. Anyway, that's my horizontal axis there.

We'll see that much of this writing - in particular in "The Technologist," below - veers more towards the individual concept of sexuality and can sometimes lose the whole idea that there's an Other participating at all. We'll also see a few examples on the spectrum where that's not true. Ok, we're going to start reading quotes and then you folks can help me map them on the conceptual matrix behind me, on the (virtual) wall.

We're going to start with a book that was cited in "The Anthology" - which I just had a chance to catch up on this morning - and it's one of the weirder examples from one of the weirder writers in sci-fi: Philip Jose Farmer (who did the "Ringworld" trilogy, among others) who really got weirder as he went... no, actually, he started weird! This piece is from 1961: "Strange Relations," essentially a novella. He's talking about a setup where a hero - a male, white hero - falls for and has a heated love affair with a buxom, gorgeous, white female... who also cooks very well and is a great wife (he makes a big point of that; all very quotidian). But he finds out - surprisingly enough - that she's actually of an insect species which evolved on another planet to perfectly mimic human physiology. The question is: what's the moral implication of the fact that he's had sex - and fallen in love with - an insect... albeit as a perfect replica of a human being? We find this "is it really an Other - or not" question, and we'll going to hit that quite a bit. The quote is discussing that general concept; in it, he says:

"Yes, you have, with a thing that is even lower than a beast of the field, the insect, what even Moses did not think of when he forbade union between man and beast, what even the forerunner could not have guessed when he reaffirmed the law and set the utmost penalty for it. 'You have done; you slept with an insect.' "

It's the crisis in "Strange Relations." The crux of the novella is: does she really "count" as a being, or not? Should this character be completely humiliated by the fact that he has had carnal intercourse with something that he really thought was a human... but it turns out otherwise? One more from the same novella, and then we'll place him on the chart:

"If you knew she were of an utterly alien branch of the animal kingdom separated by millions of years of evolution, barred by her ancestry and anatomy from the true completion of their marriage (children), you would turn from her in horror."

That belief must have injected darkness into even her brightest moments, because she was trying to hide the fact that she was an insect from him all along. Then, when he finally finds out that she's made of chitin and such, she breaks down and cries and says, "I knew you'd hate me if you found out I was a bug." He does come to hate her, actually. The whole novella falls apart: she dies, and it's all horrible... because she was an insect. The moral of the story is: don't sleep with insects, eh?

This is written in 1961, an early example. On our chart here, where do we put this - on the vertical scale, between conservative and open-ended social thinking? Pretty far down on the conservative side, eh? Farmer - for all of his weird writing - is rather conservative in the way that he looks at this particular configuration. How about focused on the individual/self, or focused on a true Other (with the capital "O")? Self, very much! The whole question is: can we convince ourselves that this Other is close enough to us that it doesn't really count as an Other? The whole idea here is that if it's too "Other," then it's by definition bad/wrong: don't do that. Bad. Moses said that was bad... simply unacceptable.

Now, let's move forward to 1968; we'll stay with Philip Jose Farmer. By this time, I assume he's experienced quite a bit of LSD - though that's pure supposition. Well, to me it seems axiomatic but, in the event, in this work he gets quite explicit in sexual matters, so... fair warning. If there are folks here who are badly flustered by dirty words then, yeah... I suspect not, but I did want to make note.

"Image of the Beast" is the name of this book, and it's quite difficult to find; basically, it's out of print. It's rather explicit in terms of sexuality. It was written in 1968 and it's been out of print for about 25 years, and it's rather unusual in several ways. I could try to tell you the story line here, although it would take 20 minutes to begin to explain... but it involves a guy whose semen is used to power a spaceship that does transdimensional time warps to return lost travelers to their home world, etc. There's a chalice that he has to fill up with semen... and that's the conventional part!

It gets weirder after that - really 'trippy' - almost hard-to-follow weird - with fairly explicit porn in the play-by-play. We can find two examples of cross-species sexuality, so let's see where they fall on our matrix. The first example, quoting his protagonist:

"I went up the hill and looked down in a hollow, and there was a female wolf, her head in four nooses, and the ends of the nooses tied to trees. She couldn't go back or forward, and there was Glam all his clothes off except for his socks and shoes [which is just a weird picture], holding the wolf by the tail and fucking her. Glam, that animal Glam, was fucking her."

Glam is this half-human, weird, super-alien thing. Part of what Farmer is trying to emphasize is that he's so half human that he's actually raping this wolf. What stands out is not the cross-species element, but rather the ugliness of the rape. It's bad, and it really shows how bad Glam is: rape as a strong signifier.

Later, the protagonist finds himself in an encounter with Glam as well - while he's sleeping with somebody who he thinks is his wife. Again, another Farmer classic here. This is what happens:

"The white body beneath him became reddish. The smooth but wet, slippery skin was covered with hairs as red as an Irish Setter's and as wet as if it had just climbed out of the water. The face became elongated and snouted. The long head-hair shrank to a bristle. The eyes shifted towards the sides of the head. The small and delicate ears became large, hairy, pointed organs. The long-fingered, well-manicured hands of his wife (that he thought) became paws with blunt-hooked nails. The legs on her shoulders became hairy, and a big hard penis was against its body. It was spreading jism over his belly and down onto his own cock, which was buried to the hairs in the hairy anus of the creature. It was too late for him to stop. He exploded inside the red-haired ass of the creature."

There's like hundreds of pages more... and this is the _milder_ stuff I'm quoting! Anyway, this is an interesting example in which he thinks he's sleeping with his wife but - instead - he's sleeping with an alien who turns into a... what sounds like a rather handsome (and well-endowed) Irish Setter, to be honest. Nice hair. Smooth lustrous coat... the whole works. If you envision the way they are actually coupled, it's interesting, eh? Unconventional. Though, to be honest, Farmer does have a fairly homophobic thing going on: gay panic, ca. 1968!

Ok, where do we put this excerpt within our vertical dimension? Different than the previous one, or essentially in the same mapping?

Man 1: Definitely in the "open" category.

Douglas: Yes, getting a little more open; it's exploring ideas - with a little bit of holding its nose in disgust, perhaps. But, really _exploring_ in tangible detail: wanting to make sure the "hairy red anus" gets mentioned twice, eh? Farmer remains really focused on such matters: there are pages and pages more of similarly-graphic descriptions, later on.

Yes, he's sliding more towards thinking in an "open" context, viz. maybe this is something interesting. Still, there's not truly an "Other" here. This creature is pretending to be human and then loses control of him/herself, morphing into an animal in mid-stride. The narrator can't stop himself, because he's too far along - and it's essentially an Irish Setter with whom he's having sex... but only by 'accident.'

Farmer is conflicted on the subject, I think. In most of his work, he never really delves deeper conceptually than this. He just gets stuck - ditto with his whiff of homophobia. To him, the "gay thing" seems assumed innately bad... but very much interesting. He doesn't go past that: bad, and interesting. Both.

Ok, next let's take a look at Jack Vance. We've got two quotes from Jack Vance. Early in his writing in 1950, we have "Tales from a Dying Earth." In it, some mendicants initiate a long religious pilgrimage. They get lost on the way home, and can't find their way back to where they were supposed to be headed. Wandering, until one group of these mendicants splits off, and...

"[w]ith the briefest of farewells, they marched to the village of the lizard folk [this is where there these huge lizard-style people, really more lizard than human in looks and behaviors], where they slaughtered the males, filed the teeth of the females, dressed them in garments of reeds, and installed themselves as Lords of the Village."

Interesting, eh? The mendicants just up and abandon efforts to get back home. They had walked through a village of lizards, and they decided "well, you know what? If we just put clothes on them and act like they are people, then that's fine". Again: "[t]hey settled down and became Lords of the Village, with their wives as lizards." Not very nice to the males, however - and what about same-sex folks? The gay side is not there - it's still 1950, remember?

True to Vance's form, this outcome is just accepted as OK; it's a part of life in one of Vance's imagined worlds. They settled down with the lizards as their wives (after filing their sharp teeth, of course!), and made a life out of things from there. Did they love those lizard-wives? One imagines they did, or they came to love them... or did they? Never explained - left to the reader to interpolate... very Vance. Indeed, if you've read from Vance's work it tends to be pretty "out there" - there's often bits of strangeness that float through the narrative, neither explained nor explored deeply. For him, it's a strange universe... things happen; don't be terribly surprised by it. This is quite different from the Farmer perspective, above, isn't it?

"Tales of a Dying Earth" is very much typical of that. Where do we put that on the horizontal vertical spectrum? Open-minded, or conservative in its general perspective? Kind of a little of both, eh? Conservative in that they try to make them look like women... but pretty open-minded in that they're lizards. I mean,

let's be candid: you can put reed skirts on them, but they don't change their bodies. They're lizards. He describes them in earlier pages: very lizard-y... scales and whatnot. Not human.

Simply put, they just gave up even pretending. It's like "well, let's just put lipstick on her and call it good!" Vance really doesn't make any big stink about this like, "oh, my God! How terrible!" Rather, he suggests implicitly that of _course_ it's what you would do if you were lost: you would marry the lizards and have lizard kids. Q.E.D. No worries. His universes tend to be just like that. Sort of "things happen, you roll with it, it's surprisingly interesting..." and life goes on.

Our next example stays with Vance, but now we're in 1979 with the "The Demon Princes" - a true classic. In it, Vance spins this story of a planet where there is a sort of a species that diverged from humanity a long time ago and has become a feral, quasi-human, separate species. Commenting on what happens to the children of these ferals, he says:

"They leave babies lying out on the fells. Simple forgetfulness, I suspect. Sometimes, they are brought in and trained, more or less successfully. Catch them early, and they'll learn to stay clean and walk on their hind legs. Tiptoe [a young female version of this] here is a clever one. She serves beer and fluffs pillows and - generally - behaves herself. 'She's fascinating to look at' says Gerson. 'Is she, well, affectionate?' 'It's been tried with, generally, poor results,' says Manoth. 'Are you curious? Touch her.' 'Where?' 'Well, to begin with on the shoulder...' "

Vance just leaves it at that. He doesn't actually take that interaction any further, although it's kind of an interesting interaction. It's sort of human... but sort of not. And it's sort of suggested that there's some cross-breeding going on here - suggested in that weird, elliptical, Vance way.

There's more of an "Other," I think, in this quote. These are really truly Others. This scene quoted above is included towards the end of this book, suggesting that maybe they're bonding with these Others as sexual/emotional partners, or lovers, or..? It's not enough to be precise. Essentially, Vance doesn't really say anything specific - though he also doesn't pass judgment against. Rather he essentially throws this in as merely "one of the things that the natives do on this particular planet."

In doing so, I think he's drifted a little more towards an open-ended conception of such relationships. But, perhaps more importantly, he's getting little deeper into a conception of an Other as a genuinely sexual being - one that really is an Other. We're not dealing with something that's merely another form of a human... one with fur on it, or whatever. This is a distinct being, a distinct species - one with its own sexual framework.

Ok, next let's go deeper into this, both temporally and conceptually. We're up to 1984 and we're quoting from "Across the Sea of Suns," by Gregory Benford. This is one of his classic, relatively early works. Our quote is from a discussion that's taking place about how to trick some aliens into falling into a particular trap... one involving black holes and quantum this and so on, and so forth. One of the protagonists suggests that maybe they can use a false neutrino beam lure, because the aliens are thought to have a sexual attraction that is mediated by neutrino beams - in a way that actually kind of makes sense if you read the whole book. Indeed, Benford's really obsessive about making sense more broadly.

So, one of the characters next states that "I never speculate on extraterrestrial pornography." This strikes me as really interesting concept. When was the last time you really thought about profoundly non-human-centric "porn?" Not like when we're putatively abducted by aliens, and UFO believers are always imagining anal probes; it's always the anal probes. OK. We've got it. Anal probes. Enough!

But wait: what do the aliens look at, porn-wise, back on their home planet? After all, we are already thinking of them as sexual beings - as sexual species

- with anal probes, right? We must imagine they themselves have some pornography, which is of course not just human porn. It would be alien porn, right? Because they're aliens, it wouldn't be visual... or would it?

This actually ties into questions that I get asked often in another part of my life, wherein I train and promote and ride show jumping horses in competition, to the International level. I work almost exclusively with stallions nowadays, both for financial reasons and also because I get on well with them. We do well together, in sport. I also do limited consulting work with folks who are having problems managing the breeding side of their stallions' careers.

A stallion won't do "stallion things," is how the conversation usually begins. "He won't do the... you know?" I ask: what's 'you know?' "Well, he won't do the 'thing' with his 'thing.' " For fuck's sake! We're adults. Can't we speak clearly about this stuff? So often, the solution to the issue when the stallion "just isn't into it" ends up being really simple: it takes basic empathy; basic caring; compassion; trust.

Alas, that's so difficult for some 2-leggers to under- stand, let alone implement. I've had people actually wonder if "we can just blow up a picture of a mare and put it on the wall - for the stallion..?" NO! Because "stallion-centric porn" doesn't work like that! That's a human thing, and stallions - by definition - aren't humans. No super-size pinups for them, no matter how alluring the mare (or fellow stallion!).

Of course, we can extrapolate what could work, right? Does anyone want to guess? If stallions could go to the store and buy "stallion-centric porn," what'd it be? Bingo: smell! It's not visual. Same with dogs. Not movies; they don't care about that. They'd have smellies... right? They'd look at our porn and think it's useless, just as smelly-porn would be useless for us.

And so, back to the issue of extraterrestrial porn: an entirely different sort of Other than mammalian porn. With it, we'd have to imagine it's basically two more strides further out from our own, shit-flinging primate conception of what pleasurable porn experiences entails. It'd be... I don't know: neutrino beam porn? This is quite fascinating, because we are now thinking of genuinely Other sexualities which are in no way mammalian - perhaps not even animalistic, in the sense of Earth-based, or carbon-based, or whatever. This brings to mind a quip from Alastair Reynolds in which he's describing a tangle of complex, world-scale machinery as akin something that a planet would enjoy as visual pornography... if planets watched porn. A clever way to say something with deep epistemological content, I think.

If we're truly thinking about a sexual universe, then other species surely must have their own quirks, eh? Trying to understand them and their particular quirks requires a big cognitive leap. That's because we take for granted that there's a certain flavor of sex that's "sex." But, what about neutrino beam sex? Can we really make sense of that? We have to be relatively clever - intellectually and empathetically - to be able to get our hands around that. I think Benford is poking at that question ontologically, and in an interesting way.

We'll hopefully get back to Benford a little bit later, but now let's move forward. Next is Stephen Baxter, in a collection of short stories published in 1997 called "Vacuum Diagrams." He's a hard sci-fi writer trained as a PhD astrophysicist. Certainly a smart guy, and in this work he's talking about a character that has been stranded in time and space... way out in the middle of nowhere.

The planet on which he's stranded, a nice person there observes that "...you are now stranded, Rhodey. You have now lost your family [because he's stuck on this planet without a spaceship to get back to his home planet]."
He says, "Yes."
She says: "You are welcome here. You could join my sexual grouping. The surgery required is superficial."
He replies: "Thank you, but that's well beyond my resources."

"Well, what then?" she asks.

So, the frame here is you've got a human who is on a planet of other kinds of humans, but it's been thousands of years that these groups have been diverging from each other. They have these groupings that are almost like new genders... but not. Rodney could be "modified" to be part of this new world of sexuality, after which he could participate in their social rituals as an equal - but he doesn't have enough money to do it. And so, the constraint is surely not moral at this point. It's rather the quotidian issue of "affording" a different sexuality, a different form of gender entirely. You can do it... but it's expensive. It'd necessary for him to fit into this culture completely... but it's rather expensive.

For me, this is a key step the literature takes here in imagining that sexuality is a component of social organization. Plus, to become truly part of a new culture, you'd need to become part of their sexuality, as well. How would one change one's sexuality? We talk about transexualism and the gender issues in humanity, but what about changing _species_ sexuality? That's a bigger jump. Now, if we want to talk about machine sexuality, with which I'll try to close within a bit, that's an even bigger jump. Because, what if we teach our machines to think in sexual terms? How will they interface with us, and where will they come from as sexual beings? Or, do we curse these singularity machines with no gender whatsoever... making them all "its?"

We now close with Greg Egan, and excerpts from "Schild's Ladder." In these, he's is talking about some characters that are conceptualized as so very far apart - in terms of diverged humanity - that one side doesn't even have bodies: they live purely in logic-space as software; the other side is still physical. The non-corporeal protagonist approaches the meatspace, and...

"He wore a look of such deadpan innocence that Cass felt sure that he knew exactly how it sounded in translation. If that was his meaning, the idea wasn't entirely absurd or unwelcome. She'd grown fond of Ramsey, even if he'd never been quite as solicitous or as eager to understand as Darsono [her same-species partner]. That truth made him all the more intriguing; if they could find enough common ground to become lovers, it might be a fitting way to bid Mimosa [the planet] farewell, sweeping away the mutually distorted views they had of each other."

So, here we have an imagined opportunity for a sexual dalliance that spans a wide spread - a gaping - chasm, between essentially two distinct species. He's also proposing that such a coupling would be a nice way to solidify a final bond, as well as a nice way to take some of the lack of mutual understanding (an absent reciprocal empathy, perhaps) and make it go away, using sexuality as a lubricant. Well, perhaps that's sort of a bad term, but... how about as a social lubricant - in order to bring people - two species of people - closer together. Using this sort of chasm-spanning sex purely for social value, in other words. The narrative continues on...

"Cass said haltingly, 'No, I'm sorry. I can't join you.'" And then she thinks to herself, "so much for feeling smugly unshockable, for daring to contemplate cross-modal sex."

This is, in a very real sense, one step beyond cross-species sexual liaison: it's cross-modal, because one is software whilst the other is physical. They'd have to build some kind of a bridge between the two to have any sexual interaction whatsoever! Cass is trying to think of herself as 'au courant' - a modern being in this universe. As someone who is willing to consider cross-modal sex - but, in the end - it just seems a little bit too challenging... perhaps a little bit too far out for her to embrace.

Now, with the above as our final quote for today's presentation, where do we place it on our 2x2 matrix? Is this a conservative or open-ended, stance, in terms of the politics of trans-human sexuality?

Man 1: It's open-ended.

Douglas: I agree; it's about as open-ended as you can get. There's no moral questioning about whether cross-modal sexual interaction makes sense or if it's in some sense morally compromised. Rather, it's in the lines of "if you can do it, then of course it'll help you understand better." That's a key piece - transcending the petty morality of a hermetically-sealed species bubble... seeing sexuality as a crucial tool for achieving truly reciprocal understanding, a reciprocal social empathy.

Next, let's ask ourselves is there's a truly other "Other" in this proposed interaction? I think so, personally. I've read the book, too, so I have a little bit more background on Egan's overall conceptual framework. In this deeper framework, these software-based human beings have been - for tens of thousands of generations - living as code. They have established a concept of sexuality that is so different from baseline human - or even animal - sexuality that it's not even cross-species; it's cross modal. It's a software/hardware interface.

This deeper questioning thereby stretches our concept of sexuality into an interesting place. And it's this expansion, this line-crossing between animal/animal sexual bonding (in the sense that humans are animals, and other mammals or other critters are animals as well) into overlapping with questions of animal/machine sexual dynamics. This is where I'm going with my deeper argument, in sum. To speak of cross-species sexuality is to speak of cross-modal sexuality - via what is simply an alternative frame of reference. The two issues are one question being asked two different ways; they're two sides of the same coin.

Sexuality in the mammals - and we're all mammals here, lest we forget - fulfills a huge number of social functions. Reproduction is one of them, but it's not the only one; this is not unique to human beings, nor unique to primates. Dolphins famously barter sex back and forth routinely between individuals for favors (i.e. back scratches) - or just as quid pro quo. Same gender, opposite gender - it varies. Thus, it's very common in mammals to use sexuality to build social bonds, and to keep people from being mad at each other. Sexuality ends up being really important.

In respect to a comment that was made in the previous presentation about the limbic system - where a lot of aggression "resides" - and if we could only dampen the limbic system, we'd probably have less aggression and hopefully less wars (because we're such a warlike species).

It's an interesting comment because, first of all, there actually isn't such a thing as a distinct "limbic system" in the mammalian brain. That was an old assumption, built upon the false emotion/logic dichotomy (see D'Amasio's "Descartes Error" for a full critique). It used to be (in the 1950s and '60s) that there was an emotional brain and then there was a rational brain - they then interacted with one another. The limbic system was supposed to be the emotional brain, with the neocortex and PFC as the rational brain.

As the neuroscientists have done more FMRI mapping and so on and so forth, it turns out that there is no distinct limbic system to the brain. The emotional components of the brain are intertwined with the other components of the brain so much that you either have to extend the concept of limbic so far that it becomes silly... or you have to just say that there's a brain.

Man 2: I think it's been referred to as the "reptile brain."

Douglas: Yep, people talked about the "reptile brain" in that context, historically.

Woman 1: There's no "reptile brain."

Douglas: Correct: there's no "reptile brain" hidden at the "bottom" of the mammalian brain, as such. Brains are an evolved, complex, interconnected bit of biological hacking and these crude attempts to split them into reified binaries - reptile versus human, emotional

versus rational - are oversimplification to the point of folly.

Woman 1: Ok.

Douglas: I actually got clued into the limbic issue by Dr. Jaak Panskepp - a very well-respected, well-published neuroscientist - with whom I talked a few years ago. He was the first one to tell me, "Douglas, there is no limbic system." "Whoa. Really? But it says in the book... I have a book right here. It says limbic..." "No, we were wrong. There's no limbic system. It's everywhere."

The "limbic system" is all everywhere in the brain - which is to say nowhere at all! We _are_ our emotions, and if you read D'Amasio - mentioned above, another neuroscientist... actually a neurologist - you'll see the alternative (and correct) epistemology he presents in "Descartes' Error." At core, he is making the argument that you cannot take the emotions out of our brain - put them into a separate "system" - be left with a purely rational brain as a result. It's a false dichotomy; we _are_ our emotions!

Indeed, one cannot really imagine us as a consciousness with our emotions pulled out - because our entire rationality is built on our emotional framework insofar as it structures what matters to us and what doesn't. Without emotionality, we are sort adrift; if anybody in here is a neuroscience geek and remembers the Phineas Gage example? He's the guy who got a rod through his head, and he lost his emotional interconnections to his cerebral cortex - thus lost his ability to make decisions because he didn't care anymore. Without emotion, it doesn't really matter.

So, the limbic system is a false phantom of human brains - mammalian brains, more generally. When we think about ourselves as these conscious beings that are struggling with emotions... that maybe we can leave such emotional baggage behind as we go forward. Such is badly-warped thinking, which doesn't at all reflect how our brains have evolved, how mammalian cognition works, how consciousness itself is structured, mediated, and implemented in wetware.

The truth is that - as is the case with every other social species of which we're currently aware - our emotions are critically interconnected with our social framework, and with our social capabilities. If we take the emotions out, we don't have a social species anymore; there is no "us" without emotional context, emotional import. The limbic system is everywhere.

Which matters a great deal, insofar as sexuality is the centre of so much emotional interaction and of our emotional landscape in general. You can't really think of us without sexuality - not to hat-tip too heavily towards Freud... but it's true. We can't remove sexuality from emotional considerations without taking so much of our emotional landscape out: essentially you've then taken out the social landscape, which means you're left with not much at all. Some of the critiques of the post-singularity stuff I've tried to make have really emphasized that importance of such emotional frameworks.

Johannes: I think we should open up the audience... that sounds strange.

Douglas: With social lubricant?

Johannes: And be gentle to the microphone, it's so poppy.

Man 3: OK, I'll be gentle, microphone. So, I wanted to ask if you have looked into much of cross-species sexuality that's actually happening, not just in science fiction. So, for example, partly because of - or inspired by - Annalee Newitz's talk two years ago in which she was asking "can we imagine sex toys that would let us have sex like octopuses?", I did this year-long project asking if we could replace the one-year requirement of real life experience that transsexuals face with one year of Second Life experience to get species change surgery. And in doing so, I discovered there is a large community of people that are very seriously trans-species: they feel on a daily basis - painfully - that they

are actually a horse or a fox or something else. So, we could look at that as trans-species sexuality, right? Or cross-species.

But also, in Second Life it's really common to be a Neko. If you saw our performance last night, our avatars are cross-species: like half human, half cat. And those are the ones that we use. In one sense, there's people who are really serious about it, and it's a really common thing because, well, it's fun. There's this whole complex interplay of species in Second Life. There are all kinds of species of ears and tails you can buy, with all kinds of different attachments... and all kinds of different genitalia for different species, and tentacles and whatnot.

And then, on top of that, there's these complex social rules. Like, there's some cities where Nekos are allowed, but furries are not. So if you're an anthropomorphic animal, that's not OK. But if you're half human and half cat - if you're like a cyborg cat, that is... possibly futuristic - then that's OK. But if you're cartoonish, then you're out.

Douglas: A couple quick comments on that. First, the world of people who feel like they're trapped in the wrong species' body - to some degree it's a furry world - though not exclusively furry. I'm not really sure we have a good name for that particular gestalt. It certainly does exist, and those folks make strong arguments (from what I understand), in a way analogous to being in the wrong-gendered body (as I believe transsexuals feel).

Although to be fair, I'm not as familiar with the literature here and thus I'm not qualified to really talk intelligently from the academic perspective. Additionally, I'm not firsthand familiar enough with the community to really talk to the core of the existential side of that position - except perhaps to confirm that it is out there. I've also been exposed to those people who would desperately love to transform permanently, very much in the same way that gender transformation is a goal for some folks because they just feel like they're in the wrong body with respect to their self-identified mental/existential gender.

Second, there's a completely different group of humanity - roughly 1-2% of human beings, across pretty much all known cultures and throughout recorded human history - who have had deep emotional, social, and physical bonds outside the human species. This has been true for as long as humans have been humans, according to all extant records available to research. It is simply accurate to say that this kind of profound cross-species intimacy (which I refer to as Deep Symbiosis) is as human as consciousness itself; this minority gift has always been with humanity, both as a discrete sexual orientation and as a fundamental form of cognitive & social organization.

Sadly, it has been attacked and pushed down aggressively in many cultures - particularly in more recent cultural systems and within monotheistic religious frameworks. Despite these seemingly-endless cultural persecutions and witch hunts, this orientation remains vibrantly visible worldwide. That, in turn, suggests that it's got relatively deep evolutionary roots: it has been highly conserved through evolutionary process, despite the social intolerance it has often faced. This is a deeply human capability - a minority gift within the population, admittedly, but centrally important to humanity's role in the wider living world.

Incidentally, the 1-2% statistic is relatively well-established data, which has come out of research efforts during the past 10 years or so undertaken by Dr. Hani Miletski and other researchers. This is a hidden (but-ubiquitous) world of folks that live and breathe and get up in the morning and go to bed at night enmeshed within deep symbiotic cross-species bonds. It is - not incidentally - a world I know very closely (to put it mildly); it also represents an interesting, challenging, largely open-ended rebuke to many of the assumptions of human uniqueness and superiority - although I don't want to bog down the present discussion with an excess of focus on this topic alone.

I think the existence of this Deep Symbiosis gift illuminates crucial elements of humanity's progress forward in dealing with true "Others," of a machine sort - as well as Others of an other species sort - because we already have humans who interface physically/emotionally/sexually with non-humans in a productive, healthy, mutually-reciprocal way. To me, that's a genuinely interesting species skill to find present, very much so!

I argue that it's in our genes; that it's expressed rarely, yes, but it's there. It's part of who we are as a species. It ties into our empathetic capabilities, and it's also tied into autism in interesting ways... yeah, that's a couple of days' worth of presentations and discussions, in themselves. There's a lot of published literature - some of which I've authored - on this, if you care to follow up.

Man 4: Thank you. I wondered if you could explain why you chose to plot these different science fiction authors and their works on what I think most people would recognize as the political compass, and how you think the need to plot it on that dimension influenced your choice of science fiction authors.

Douglas: Good question. Actually, it went the other direction: I pulled all the quotes that I could find which were meaningful, and concise enough to stand somewhat independent of narrative flow. Those quotes were pulled from several hundred books... a wide spectrum, collected over the years. Such examples are not plentiful, to be honest.

What happened is this: I have been winnowing the literature for years, selecting quotes, and sequestering them in a little corner of my computer. Finally, I went through them, pulled out all the quotes: I read through them, and tried to craft a useful taxonomy into which they'd fit.

My first instinct was to scan for an historical progression... you know, start with the old stuff, and move to the new stuff. That didn't really hang together, a linear historical progression. After that, I tried out this 2x2 matrix - mostly as a place to throw darts on the dartboard and test out these different dimensions through which cross-species questions can be mapped. But they're not the only two spectra that I think are important, and thus are somewhat arbitrary at heart.

I think a centrally important finding is really that there's so little insightful cross-species material present in this literature. It's amazing to note, when compared to the fantasy literature. And - keep in mind - in sci-fi there's a lot of cross-species frissage going on. I mean, there's always aliens... they're hanging out, probing, etc. Take Alastair Reynolds, his Revelation Space universe: it's what... 5 books now? All sorts of aliens, including aliens who are entire oceans... planets! ...and the "sex" word really just like doesn't come up. Not one cross-species intimacy example in 5 books. So, it's somewhat of a hidden question, one that science fiction rarely touches. But... when it does, it always has interesting things to say.

Johannes: Yeah, there's actually - in the new Arse Elektronika book - which came out like three days ago... well, I have to suggest that... I think that people should read it, OK? So, there is a story by James Tiptree, Jr., who was actually female. It was a pseudonym of Mrs. Sheldon. And she wrote in 1972 this really exciting short story.

Douglas: Additionally, there's terribly relevant quote from Rudy Rucker; in choosing it, I'm also pimping for the aforementioned Arse Elektronika 2008 anthology! He says that...

"[o]ne reason we're attracted to sex with other people is simply because they're different. Gender isn't necessarily an issue. That's the core idea in both the Delaney and the Sheldon stories [which he is reviewing]. Otherness is a turn on. And any other person is, for all practical purposes, an alien... if you really think about it. Note that it's not just the difference that turns us on, it's the idea that there's an "intelligent" mind inside a different body. Another mind that mirrors you. A mind you can, in fact, pair

up with for endless regressive mutual reflections." [emphasis added]

I think Rucker truly hits the nail right on the head with that. I mean, the eroticism of the Other is real - and that's a big part of sex... of healthy sex that brings together beings into something more than just a physical compilation of rote behaviors. Any other kind of sex - without a genuinely "Other" other - is really a form of masturbation (not that there's anything 'wrong' with that... but it's not sex, eh?).

Even if we're in a same-sex couple within our own species, there's still an Otherness there that is deeply attractive, deeply mysterious. That's what truly makes it interesting: the unknown, the parallel-but-different sentience, the mystery embedded in all forms of genuine empathy, the reciprocal trust and respect. We have a word for all that, don't we? It's what we talk about when we talk about "love" (with apologies to Ray Carver!).

And the intelligence behind the Other matters... it matters a great deal! A lack of intelligence - of consciousness - of embodied sentience in a sexual partner turns an event of mutual, reciprocal intimacy into an act of one-sided pleasure; again, not intrinsically horrific... but it's not love, it's "just sex" or it's whatever monochromatic substitute we want to call it in the English vernacular.

In the event, I think - and I hope - we're going to see a lot more substantive exploration of these ideas - the overlap between the cyborg and the cross-species conceptual frameworks. We'll see work that continues blurring lines - which continues to question the epistemological assumptions of such lines in the first place! - as science fiction tries to get its hands around thinking about Others as genuinely social entities... not just as foils for ourselves. We're past the alien as childish stick figure with a big head metaphor; it's time for aliens with the full spectrum of social complexities, personal attributes, and existential challenges... past time, really. We're blessed to be surrounded by such "aliens," here on our own home planet, in the form of nonhuman species - and yet we have hidden from their reality, shirked our capability to know them and appreciate them and see the world through their senses. It's a terrible shame, isn't it?

In the event, it's through the genuine extension of human mental capability into something that's not simply navel-gazing (philosophy) or toy-playing (physics, chemistry) that we will find our most fascinating and most challenging cognitive puzzles, nowadays. To understand - to truly appreciate and come to empathize with - intelligent Others (be they other species on our planet, non-biological post-Singularity entities, or aliens from other locales)... that's the genuine frontier of human cognitive achievement.

We can do it; we already have the demonstrated capability of Deep Symbiosis - and the presence of such gifted individuals throughout history. Their existence shows us we have that capability, definitively so! But, to do so we must transcend our own limited cognitive frameworks, our own self-important assumptions, our own myopic expectation that the only yardstick for measuring the rest of the sentient universe is a blurred reflection of ourselves.

Once we've done that, we are truly at the next phase of our species' cognitive evolution. Whether we genuinely embrace those steps forward - or are instead pulled back into a brittle, self-centered, blinkered, hysterical worldview that measures everything through limited human terms and a tiny lense of purely human preconceptions... this is the great fork in the road facing our species today. Which fork will we follow?

The Singularity - in its own way - brings the question to the forefront, in parallel with the essential cross-species critique at which I've tilted today. One way or another, we can't keep hiding from this decision point: are we merely human, or do we celebrate our ability to understand, to embrace - indeed, to _be_ - more than ourselves?

Thank you for listening, and for honoring me with your interest and questions. Goodnight.

Johannes: Big applause for Douglas Spink. [applause]

Ani Niow

STEAMPUNK VIBRATOR

Why...

...create a device such as this? It's impractical! It'll burn you! It weighs too much to be truly useful! It's not really steampunk!

Here's the thing about art. It is often not practical and not made for general consumption. When I asked my friend Alex about his target demographic for his Tactical Corsets he replied with *"My target demographic-is me."* Despite this conversation happening months after creating the Steampunk Vibrator, it applied to my whole thought process behind it. My target demographic was indeed me, I created this device for myself first, and then it resonated with the rest of the population.

That said, it wasn't entirely unintentional that it received the attention that it did, though the scope of it was completely unexpected. Especially as I had not submitted it to any of the sites it ended up on. It was completely and totally viral.

One day I made a tweet and it got re-tweeted. And re-tweeted. It got picked up on *Laughing Squid* and then exploded. I woke up one day to have too many new random followers to count and more hits on my *Flickr* than I've ever received before. As I mentioned, it wasn't completely unintentional that it received at least some attention. One of the purposes behind this was a bit of publicity as I was unemployed and looking to get at least an internship. I figured making something as radical as this on my first try in a machine shop would prove I have the skills and creativity to be employable. I did fulfil that goal though not in the way I had anticipated.

This isn't the first steam powered vibrator to ever exist, the earliest I've seen was called the *Manipulator*. Those who think that my device was impractical have never seen or heard of a vibrator that was so big it took an entire room to house it. I've never before seen a steam powered handheld one though.

In my machine shop class at *City College of San Francisco*, the official project was to create a tiny twin cylinder steam engine, known as the *RV-1*. I had an epiphany one day that I had to make that run a vibrator somehow which naturally got me way more excited about the project. What else could I possibly be using a tiny steam engine for?

That's where I ran into my first problem. How on earth am I going to make a vibrator run off of that thing? It's not powerful enough to run anything except itself, even then barely so. I really wanted to put the engine inside the vibrator itself if only to see if it could be done. Making something that small and powerful enough would also seriously test my skills, especially as I've never worked in a machine shop before this.

Salvation came when one of my classmates was working on an engine I've never heard of. A Tesla Turbine. It looked like the perfect solution for my project. High RPM, compact, and not all that difficult to build. Not to mention is has Tesla in the title, making its steampunk cred more legitimate.

In summary, the Why? can be described in one short paragraph. I wanted to create a handheld vibrator that I could potentially run off of steam. It is an art piece, a proof of concept, a test of my machining and design skills and a somewhat unintentional publicity stunt. It is not however, a practical vibrator. I knew this would be the case if I were to design it the way I wanted to. Steam will burn you and air compressors are not sexy. It works absolutely fantastically as a, rather heavy, metal dildo, however.

How...

...did this device come into the physical world? Unlike most vibrators that spontaneously appear out of factories in China, this one was handmade in a machine shop at a community college in San Francisco. On big machines that are meant for big projects, not small ones like this.

When I got the basic concepts in place, it was time to cut some metal. Since this was supposed to be "steampunk", I wanted to make sure it didn't have any plastics, silicone, or fancy polymers. Alternate universe interpretations of Victorian era Western culture had none of these, but they were all big on metal. Not all types are safe to use in the sex toy sense, however. Non-anodised aluminum doesn't always react well to wet environments, regular steel will rust, and some people are allergic to brass. I then thought of what metal toys already exist and they were all anodised aluminum or stainless steel. I didn't have the equipment to anodise stuff there at CCSF but I could certainly machine stainless steel on these lathes.

If you look closely at the base where the engine is housed, you see multiple layers of steel. This had three purposes; 1) it gave tactile feedback 2) it enabled me to trim off layers at a time once I got closer to the size I wanted and 3) I liked the way it looked. I wanted the external design to be relatively simple, but also elegant.

The base is hollowed out for the engine with a threaded entrance for the engine cap. There's also a threaded hole drilled into the side to make way for the hose adapter.

Speaking of the engine, here comes the really fun part.

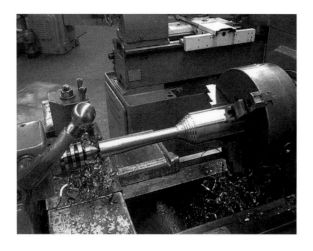

I started off with an approximately 14″ long 2″ thick piece of stainless steel, cut it in half, and machined one of those halves down to what it ended up becoming. I ended up with lots of stainless steel wool as a result.

In true steampunk fashion, there had to be brass in the design somewhere. Since it was not safe to put it in a place where it would come into contact with skin, I decided to make it internal.

This is actually version 1 of the engine housing. I discovered Tesla turbines do not work like impellers and prefer to create a spiralling effect to move fluid from the inlet to the exhaust. The current version has only one of these slots. Those were the only part of the engine that was created on a mill instead of a lathe.

As with most practical vibrators, the engine uses an eccentric weight on the shaft to create vibration. This took a few versions to get quite right as I discovered just how low in torque the Tesla actually is. Even a little weight like this slows it down significantly and I had to make it small enough, but heavy enough, to be optimal for this purpose. The one pictured here is actually quite a bit larger than the one it currently uses.

This is an old version of the engine, but what's notable about this pic is it shows the actual rotors for the turbine from the side. I decided to use tiny diamond coated cut-off wheels with pre-drilled ventilation holes around the centre. These were originally designed for a *Dremel*. Yes, this vibrator is indirectly *Dremel* powered.

The cut-off wheels were suggested by my shop instructor since they were designed for high RPM applications and they were already the size I wanted. I did some searching and found some that had the exhaust holes I needed for Tesla's design already drilled in. And cheap! Perfect!

Tiny washers were then used to separate the rotors from each other. According to a bit of internet research I did on this kind of turbine, it was recommended that they be 0.014″ thick. I was lucky enough to find some that fit, but given the uneven nature of the rotors, I sometimes had to use two or more to get any kind of separation. If I were to do this again I would probably make my own rotors instead of using the *Dremel* ones so I would have more control over how I wanted them to be formed. Since it is such a high RPM engine I also needed small enough bearings to fit inside. A couple of shaft collars were then used to keep the rotors of the engine together so they don't become misaligned inside the housing.

The engine cap screws directly into the stainless steel housing and fits against the engine enclosure. It has a little lip that a classmate of mine made when I was frantically trying to get this done before its unveiling at *Femina Potens*.

Most of these pictures were taken before it was properly bent, a-la G-Spot style. I had originally wanted to make that curve before the exhibit opened but I didn't have time. I did however, take it out of the gallery after the unveiling, bent it using a hydraulic press, and it was done.

I had brought a pressure cooker to test out true steam but it didn't get quite powerful enough to make it go very much at all. It did however get very very very hot. So hot that I had to use welding gloves to hold it and even those were starting to get toasty. The engine is quite quirky, the bare rotors seemed to run fine with steam on their own but not inside of the housing I made for them. Compressed air seemed to do the trick so I put no further development into getting it to run more optimally on steam when it became too dangerous to handle.

WTF!? I mean seriously, WTF?

Exactly.

This entire project was one WTF after another. From concept to execution I ran into that question all the time. It needed to get done. If you've ever had the pull, the desire to create something, you realise that you can't rest until it is complete.

Is it a practical vibrator? Seriously? Something that if ran off of steam would require not only a boiler powerful enough to run it, but it would *burn* you, even with gloves. Gizmodo's article about it was right; it may actually scorch your privates. Not to mention it weighs 2.5lbs, so the sheer mass absorbs much of the useful vibration.

I never ever once thought this would be a practical device and made no claims as to it's usability as such. See the Why? section for more details. Among many things this is an art piece.

Has it fulfilled its purpose as a sex toy? You bet!

And there you have it. The Steampunk Vibrator. The how, the why and the WTF.

http://steamfuck.me

UTOPIAN SEX IN IAIN M. BANKS' "CULTURE"
EXCERPT FROM BANKS' TEXT "A FEW NOTES ON THE CULTURE"

The Culture is a fictional interstellar anarchic, socialist, and utopian society created by the Scottish writer Iain M. Banks which features in a number of science fiction novels and works of short fiction by him.

The Culture is characterized by being a post-scarcity society (meaning that its advanced technologies provide practically limitless material wealth and comforts for everyone for free, having all but abolished the concept of possessions), by having overcome almost all physical constraints on life (including disease and death) and by being an almost totally egalitarian, stable society without the use of any form of force or compulsion, except where necessary to protect others.

Here is an excerpt from Banks' text "A Few Notes on the Culture" that was relevant for the creation of the Arse Elektronika conference.

One idea behind the Culture as it is depicted in the stories is that it has gone through cyclical stages during which there has been extensive human-machine interfacing, and other stages (sometimes coinciding with the human-machine eras) when extensive genetic alteration has been the norm. The era of the stories written so far - dating from about 1300 AD to 2100 AD - is one in which the people of the Culture have returned, probably temporarily, to something more 'classical' in terms of their relations with the machines and the potential of their own genes.

The Culture recognises, expects and incorporates fashions - albeit long-term fashions - in such matters. It can look back to times when people lived much of their lives in what we would now call cyberspace, and to eras when people chose to alter themselves or their children through genetic manipulation, producing a variety of morphological sub-species. Remnants of the various waves of such civilisational fashions can be found scattered throughout the Culture, and virtually everyone in the Culture carries the results of genetic manipulation in every cell of their body; it is arguably the most reliable signifier of Culture status.

Thanks to that genetic manipulation, the average Culture human will be born whole and healthy and of significantly (though not immensely) greater intelligence than their basic human genetic inheritance might imply. There are thousands of alterations to that human-basic inheritance - blister-free callusing and a clot-filter protecting the brain are two of the less important ones mentioned in the stories - but the major changes the standard Culture person would expect to be born with would include an optimized immune system and enhanced senses, freedom from inheritable diseases or defects, the ability to control their autonomic processes and nervous system (pain can, in effect, be switched off), and to survive and fully recover from wounds which would either kill or permanently mutilate without such genetic tinkering.

The vast majority of people are also born with greatly altered glands housed within their central nervous systems, usually referred to as 'drug glands'. These secrete - on command - mood- and sensory-appreciation-altering compounds into the person's bloodstream. A similar preponderance of Culture inhabitants have subtly altered reproductive organs - and control over the associated nerves - to enhance sexual pleasure. Ovulation is at will in the female, and a fetus up to a certain stage may be re-absorbed, aborted, or held at a static point in its development; again, as willed. An elaborate thought-code, self-administered in a trance-like state (or simply a consistent desire, even if not conscious) will lead, over the course of about a year, to what amounts to a viral change from one sex into the other. The convention - tradition, even - in the Culture during the time of the stories written so far is that each person should give birth to one child in their lives. In practice, the population grows slowly. (And sporadically, in addition, for other reasons, as we'll come to later.)

To us, perhaps, the idea of being able to find out what sex is like for our complimentary gender, or being able to get drunk/stoned/tripped-out or whatever just by thinking about it (and of course the Culture's drug-glands produce no unpleasant side-effects or physiological addiction) may seem like mere wish-fulfilment. And indeed it is partly wish-fulfil-ment, but then the fulfilment of wishes is both one of civilisation's most powerful drives and arguably one of its highest functions; we wish to live longer, we wish to live more comfortably, we wish to live with less anxiety and more enjoyment, less ignorance and more knowledge than our ancestors did... but the abilities to change sex and to alter one's brain-chem-istry - without resort to external technology or any form of payment - both have more serious functions within the Culture. A society in which it is so easy to change sex will rapidly find out if it is treating one gender better than the other; within the population, over time, there will gradually be greater and greater numbers of the sex it is more rewarding to be, and so pressure for change - within society rather than the individuals - will presumably therefore build up until some form of sexual equality and hence numerical parity is established. In a similar fashion, a society in which everybody is free to, and does, choose to spend the majority of their time zonked out of their brains will know that there is something significantly wrong with reality, and (one would hope) do what it can to make that reality more appealing and less - in the pejorative sense - mundane.

Implicit in the stories so far is that through self-correcting mechanisms of this nature the Culture reached a rough steady-state in such matters thou-sands of years ago, and has settled into a kind of long-lived civilisational main sequence which should last for the forseeable future, and thousands of generations.

Which brings us to the length of those generations, and the fact that they can be said to exist at all. Humans in the Culture normally live about three-and-a-half to four centuries. The majority of their lives consists of a three-century plateau which they reach in what we would compare to our mid-twenties, after a relatively normal pace of maturation during childhood, adolescence and early adulthood. They age very slowly during those three hundred years, then begin to age more quickly, then they die.

Philosophy, again; death is regarded as part of life, and nothing, including the universe, lasts forever. It is seen as bad manners to try and pretend that death is somehow not natural; instead death is seen as giving shape to life.

Monika Kribusz

SHAKTI LILA
THE CREATING GAME OF ENERGY

A very short introduction to Tantra

Tantra is a spiritual pathway. A philosophical, religious and magical movement oriented around insight through experience. Tantra is a psycho-physical way of self-realisation. The roots of this spiritual path are based in India. The golden age of Tantra is dated between the 5th and the 12th centuries AD. In the western world well-known religions like Hinduism, Buddhism and Taoism were dealing with Tantra. It influenced cults and spiritual directions like Jainism, Bön, Sikh etc., traditions which are less known to the Western world. India, Nepal, China, Japan, Tibet, Korea, Cambodia, Burma, Indonesia and Mongolia were the countries most influenced by the Tantric tradition.

The way of the Tantric disciples in ancient times took 12 years in the Hindu tradition and was a complex system of developing the disciplined skills and abilities. Tantra taught respect for others and acceptance of our different desires for happiness. Living entities, animals, plants, ecosystems, nature, planets and the cosmos were understood as parts of the same holistic "One". It was treated respectfully and named by thousands of Gods and Goddesses. It was a spiritually-integrated way of simplicity; a life together with our resources. At the same time it was not ascetic but celebrated life in all it aspects. Tantra´s followers understood themselves to be part of a holistic system held together by the same spirit. The tantric culture contained ethical rules, yoga, meditation, visualisation, knowledge about the energy system of the body, rituals, mathematics, astrology and astronomy. Ayurvedic medicine as well as literature, poetry, fine arts and architecture were related to it; it was a movement, which involved every part of life in a society.

The "blossoming period" in India ended around the 12th Century, caused by historical and political changes. In the times of the sultans of Delhi from the 13th until the middle of the 16th Century; in the time of the Mogul Empire, from the middle of the 16th Century until the middle of the 19th Century; and in the time of British regime from 1858 until 1949;

society was not really open to the Tantric culture. Ascetic vedic-brahmanic traditions in Hinduism became more influential on society at this time. As a result, many temples, sculptures and scriptures were destroyed. Tantra survived mostly in the Hindu, Buddhist and Taoist contexts, at the periphery of the Indian Sub-Continent; in little villages in the South; in remote monasteries in the forests; as well as in the mountains of the Himalayas.

Western travellers, scientific researchers and spiritual adventurers brought the tantric knowledge - at the beginning of the 19th Century - to the West. At the beginning of the 20th Century some individualists interested in mysticism, spirituality and alternative forms of living published a few books about Tantra and the rituals of initiation. It has become increasingly public since the 1960s. It is not surprising that in the West - the sexual aspect was highlighted as Western civilisation has not had a similarly high cultivation and celebration of eroticism and spirituality since the Greco-Roman antiquity. The West did not cultivate its sexual world heritage. For long periods of history, the culture and cultivation of sexuality could not be developed. The acceptance of diversity in sexuality and the rich variety in erotic practice could not be openly expressed.

In the 18th and 19th Century, Western intellectuals, adventurers, voyagers and scholars discovered the tantric way for the western world. At the beginning it was of interest only for a few individualists, spiritual seekers and academics. Later in the 20th century, Tantra became, through a growing interest in Eastern philosophy and culture, more popular in the West. At the same time, the West was becoming more self-critical of western religions, western spirituality and the western way of life in general.

At the middle of the 20th Century, alternative subcultures became more important and courageous. The coming up of the student movement in the 1960s, the postwar pacifist trend, the sexual revolution, the countless socio-political movements, the de-conditioning of old habits and moralistic rules in society were the result. Tantra, with its spiritual, exotic and taboo-breaking character fitted perfectly to these changes. People became more open to consciousness expanding techniques and psychedelic experiences. A new contemporary form of Tantrism: Neotantra was born. In the Neotantric movement sexuality was highlighted. In the ancient teachings sexual-rituals were an integral part, but not the main part. In Neotantra sexuality became the most important component. The West had not cultivated their sexual world heritage in the last few hundred years. In the translations between ancient tantric traditions and Neotantra a lot of misunderstandings were created by focusing on the role of sexuality and reducing the role of spirituality. The new form – corresponding to the expectations of western seekers – was a hybrid of contemporary dreams and interpretations, alongside old techniques and teachings.

equality of genders; sexual diversity; the cultivation of eroticism; our connectedness and responsibility to an intact nature; and also our dependence on it.

Shakti lila – The thermal-imaging camera project

I started a pilot project to measure the temperature changes in different areas of the body during a tantric ritual. It is a pilot project, only the first step in a completely unknown field. In the future we want to continue by measuring brainwaves, hormone production of glands and organs, respiratory and heart rate etc. We recorded a tantric sexual union with a thermal imaging camera. A thermal imaging camera is normally used to photograph buildings, to see how good or bad the isolation of windows, doors and walls is. We captured the ritual, including

For thousands of years, breathing -meditation- and trance methods were developed in tantric sexual-rituals. Physical and psychological practices were refined over the generations. Consciousness expanding methods, energy patterns, hormone production etc. worked very similarly over several generations. The alchemical process of the human body works independent from era and cultural context. It works therefore, also in the 21th Century, in urban society, in the globalised world, a world that is just beginning to explore spirituality again; the

the meditations, the chants, the different ritualistic parts and naturally the sexual union itself. Thermal imaging cameras measure the heat of the body, of the air, and of the objects in the room. The temperature of different body parts changes as a natural reaction during the union. I guess it can show a completely new aesthetic and reality about sexuality – than we are used to. The positions, which require a large amount of body flexibility and an intense practice of yoga, are recorded. The bodies and the sexuality are painted in the shifting colour of the thermal imaging

camera: another reality is made visible. Because there are no measuring instruments for subjective feelings, for the metaphysics of the psyche, we measure the physical reactions.

Tantric rituals work with the alchemical processes of the body. The aim is to get spiritual insight through experience, expressed through typical eastern vocabulary; "to reach enlightenment." Including sexuality, as a part of a holistic way of life is the norm in tantric philosophy. To use the vehicle of human desires for enlightenment and insight is an integral part of the tantric approach. To eliminate the difference between subject and object – what is created in the mind, if the mind is used in everyday life – is also an aim of the conscious use of applied sexuality. Different techniques of meditation without using sexual intercourse have the same aim, by including sexuality in meditative practice one can reach these states of consciousness more quickly. Divine consciousness probably sounds a little bit unusual to us, but it simply means identification with the whole universe. Identification also means taking responsibility for everything we get in touch with – including getting in touch with people, nature etc. It means our everyday responsibility to live together in a constructive way – it is the way of humanity, sensuality and joy not only in the ritual or in hidden places out of this world but in everyday life situations. Sexuality is also a generator for more energy in life. This energy that is generated in the ritual is used for the realisation of our desires. At this point the tantric ethic plays a prominent role, because the most important desire is to live in understanding, harmony, tolerance and happiness in society.

The human consciousness has deeper levels than only waking consciousness. The deeper levels, known in general as the subconscious, determinate our actions and lives more than we consider. To reach and deal with these subconscious parts in a deep trance is one of the most important parts of tantric sexuality. The ritual has to be a good mixture of strict procedure and training, days before the ritual, but also space for spontaneity, creativity and self-expression.

Since the scientific study of human consciousness in neurobiology and psychology focuses on research, scientists want to measure changes. As far I knew the changes in the body during tantric sexuality had not been measured at all. Tantrics tell about different phenomena, about their subjective feelings and unusual extrasensory experiences; about all the perceptions and sensations that overwhelm them during a ritual. Sensations, kinds of feelings, they had never experienced before and for which we do not have any words in our culture; even some feelings which are beyond words.

Tantric sexuality is also a generator for more energy in life. The energy is generated in the pelvis and canalised through the central nervous system to the brain, specifically to the pineal and the pituitary gland, which are hormone producing glands. They play an important role in reaching different states of consciousness. Tantra uses different yoga-like positions, breath techniques, mantras, visualisations, compression, rubbing and massage of different acupressure points and energy centres of the body to grow and canalise the energy. It also deals with the catapulting power of poetry, emotions, openness, honest and serious understanding and acceptance for each other, which is most important and above any other technique.

Tantra is very open-hearted to different sexual orientations and points out the bisexual nature of human beings – we have been familiar with this thesis in western psychology since Sigmund Freud. Tantra means we have to develop the female and the male parts on every level of our personality to reach our whole potential.

Tantric sexuality is a strongly structured ritual, at the same time it develops a deep understanding for each other, a sensitive slowness, breaking of taboos and in fact a celebration of taboos. To put another way, it is about the generating and the channelling of energy, of life force, to harmonize our opposing sides and energize ourselves, to transform sexual energy and use it for creating art or intellectual work, with the ultimate aim of just experiencing how high our happiness can be.

The tantric ritual we recorded took 4 hours so we had to cut the material for conciseness. For example there is a long meditation at the beginning - it can be boring for the audience to watch two sitting people in a yoga position who do not do anything for 20 minutes.

Rituals are important, because the psyche understands the symbolism of gesture and body language much more than intellectual descriptions. In the ritual everybody feels whether the partner really mentally opens up or not. At a certain point the bodies start to speak to each other, the partner feels – as they say – like they were "moved by sexuality", like something would act through them.

Different positions in the film are taken from *Kama Sutra*, *Ananda Ranga* and *The Perfumed Garden* – old scriptures from India about how to celebrate love. A little time is needed to get the body into these positions – for example sexual stimulation while doing a headstand, which has a special effect on the energy flow of the body, especially of the pineal gland.

The film is just the beginning of a longer project to picture new and old ways of erotic culture and to show unusual methods and develop new ones of composition and depiction of eroticism. We want to encourage try new ways in sexuality and experience ourselves differently than we are used to. Sexuality is something that moves everybody, it is time to cultivate, celebrate and learn more about it how to do it in love.

Noah Weinstein and Randy Sarafan

THE JOYDICK

The Joydick is a wearable haptic device for controlling video gameplay based on realtime male masturbation. Through the use of a carefully designed strap-on interface, the user's penis is converted into a joystick capable of moving the character onscreen in all four cardinal directions. For games requiring the fire button, a separate ring can be worn which converts hand-strokes into button presses.

The Theory

The "core mechanic" is the action a player does over and over again during game play. This may be rolling dice or it may be frantically pressing a button. Although, this behavior tends to vary, the objective of this behavior is always the same, to win the game.

Our impetus to win can be seen as a drive towards transcendence; a transcendence that is both over death and, in a sense, a metaphorical death. Winning a video game is much like what Martin Heidegger referred to as becoming a "being towards death." That is a self-realized individual who has overcome uncertainty in life, reconciled their place in the universe and has acknowledged death within their life.

This simultaneity of both transcendence in life and the acknowledgment of death are also encountered during what the French like to call *la petite mort* or in English, "the little death." This is the refractory period following sexual climax in which a person can achieve no further orgasm and is filled both with pleasure and melancholy.

It would be reasonable to assert that the tension that builds during gameplay and the release achieved through victory are similar to the events leading up and through a sexual orgasm.

Bringing us full circle, aside from sharing a similar goal and end result, the much more obvious relation between video gameplay and what this haptic technology provides, an expression of masculine sexuality is that they are both driven by a core mechanic. In the case of male masturbation, the core mechanic is the repeated stimulation of the nerve endings of the penis.

In fact, the similarities between the mechanics and objectives of both sex and video games are so striking that it may be fair to say that gameplay, particular video games, are driven by displaced sexual energy.

This hypothesis can help explain why young, sexually-frustrated males are the largest demographic in the gaming world and why men in general are two times more likely to be avid gamers than women. It is also true that energy misdirected towards video games has been known to destroy marriages and tear apart relationships. In fact, Woman's Health has published an article called Video Games and Foreplay in their Sex and Relationships column in which they state that women should play more video games as a way to proactively engage with their male partner. What is striking about this is that by likening video gaming to foreplay and claiming it as a trust-building exercise, Women's Health Magazine is implying that video games, in some circumstances, have taken the role of actual sexual foreplay.

The link between video gameplay and male sexual stimulation seems quite clear. It is therefore somewhat surprising that (it would seem) up until now, no one has taken the logical step forward and used repeated sexual stimulation as a means for controlling a video game.

Although a number of people have used game controllers as stimulators by creating devices that are reactive to gameplay, a majority of these systems are directed towards female stimulation and are possibly designed as a way to engage frustrated partners in the masturbatory nature of video gameplay. Such thoughts were also alluded to by Jane Pinckard while reviewing the Japanese game Rez which is often sold with a "trance vibrator."

Given that most sexual gaming interfaces are responsive to gameplay and directed towards female stimulation, the Joydick is a substantial breakthrough and brings the first proactive male-oriented sexual gaming interface to the world.

The Joydick stimulates males both mentally and physically by combining the core mechanic of gameplay with the core mechanic of sexual stimulation.

Make your own!

You will need:

System
- An Atari and expendable joystick controller - we used the Atari Flashback 2 system.
- A phallus - we used the Flex Gordon from Good Vibrations, you can use your own.
- 5V 500 mA power supply - we used an old cell phone wall charger.
- Tissues
- Five 5V Relays (W107DIP-5)
- 16F877 Pic chip
- 20 MHZ Resonator
- A 40 Pin socket
- 0.1uF ceramic disc capacitor
- Two 10K resistors
- 330 ohm resistor
- Wire
- Female telephone handset jack
- A telephone handset cord

- A PCB (Radioshack #276-150)
- A small thin piece of wood or acrylic

Sensor
- Tilt switch - we used the one we hacked apart from inside the Cube World Dodger Game we had lying around the lab.
- IR LED receiver - Digikey 475-1441-ND
- Coiled telephone cord
- 4.7K resistor
- 220 ohm resistor
- 33K resistor
- 100K resistor
- Art Stuff SR-2330 Silicone and blue pigment
- Two part epoxy

Ring
- IR LED emitter - Digikey 516-1262-ND
- CR2032 3V lithium coin battery
- Coin battery holder - we used one salvaged from old computer mother board
- Stock ring from jewelery or bead store
- Solid core thin gauge wire

Strap
- 1 mil neoprene fabric
- 3/4" Velcro
- 3/4" elastic banding
- 1/4" elastic banding

Tools
- Rotary cutter or scissors
- Needle nose pliers
- Electrical pliers
- Wire strippers
- Hot glue gun
- Sewing machine
- Volt meter
- Soldering iron
- Dremel
- Drill
- Two part epoxy

A small dab of hot glue or epoxy is a good idea to hold the LED in place and protect against shorts.

Power the ring with the 3V coin cell battery.

Make a ring

Like The Lord of The Rings, this epic journey starts with a ring.

The ring emits IR signals which are detected by the base sensor, interpreted by the pic chip, and then sent to the game console as the fire or trigger button signal. Making the ring is easy.

Head down to the local craft or bead store and pick up a blank ring. We cut a slit in our ring so that it would fit over a variety of finger sizes. Then, using a rotary tool, grind down a flat spot on the ring - the coin cell battery holder will be epoxied on there.

Salvage a coin cell battery holder by desoldering one from an old computer mother board and attach it to the ring using epoxy as pictured below.

Then, solder a small wire onto the negative (-) lead from the battery holder and glue it into place on the front of the ring where the IR LED will be installed.

Last, cut the leads of the IR LED as short as possible and solder them onto the end of the negative wire coming from the back of the battery holder and the positive lead coming off of the battery holder so that the LED is pointed towards your pinky finger when wearing the ring on your ring finger.

Hack Cube World

With the ring (emitter) complete, it's time to make the sensor. The sensor is comprised of a tilt switch and the phototransistor IR LED's and phototransistors can be found online from Digikey (see link in materials section).

We needed a 4-way tilt sensor, and just happened to have a few lying around the lab embedded in some game modules called Cube World that we weren't currently using.

Open up your cube game and locate the tilt sensor. It is the boxy thing in the center that sounds like there is something rattling around inside.

Cut the tilt sensor away from the circuit board by trimming the ribbon cable as close to the circuit board as possible.

Break/cut the tilt switch out of its mounting points to the rest of the cube and free it from its cage. It's destined for a far greater future.

Pin 2 must be left open for a ground wire.

Attach a single black wire to Pin 2 as this is the ground cable. All four positions on the tilt sensor share this common ground cable.

Lastly, twist all the resistors together with a red wire and solder.

Now for each cardinal direction the tilt sensor will produce a different resistance. The micro controller will be able to read these resistances as different values.

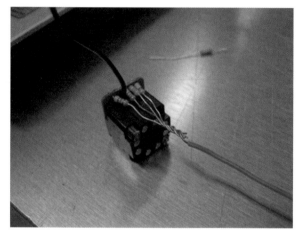

Attach Resistors

Desolder the ribbon cable from the tilt sensor. Replace the pins with resistors as follows:

- Pin 1: 4.7K
- Pin 2: No Connect (for now)
- Pin 3: 220 ohm
- Pin 4: 33K
- Pin 5: 100K

Pins are read left to write when holding the tilt sensor with the solder holes facing closest to you.

You don't need to solder them in this particular order, but this is the ordering the code is calibrated for.

Testy

Test the tilt sensor and IR LED with the phototransistor to make certain that it works with the code.

Below is the code that was used. It was written in Mbasic which is a variant of Basic designed to work the Basic Micro PIC development environment. It is pretty simple and can be easily converted to other languages.

```
CPU = 16F877
MHZ = 20
CONFIG 16254
```

```
'++ Defines' the variables ++
IRsense var word
tilting var word

'++ Sets constants for analog in ++
AN0 con 0
CLK con 2
ADSETUP con %10000000

' ++ Starts main code loop ++
main:

'++ Configures analog in pins and reads to
IRsense variable ++
ADIN AN0, CLK, ADSETUP, IRsense

'++ If the ring is close to the button it turns on
the fire button relay ++
if IRsense < 50 then
High B6
pause 100
else
Low B6
endif

'++ code to read the tilt sensor ++
High B1
pause 1
RCTIME B1,1,tilting

pause 1

'++ Turns on and off direction relays depending
on tilt sensor readings ++
if tilting > 0 and tilting < 10 then
High B5
pause 1

else
Low B5

endif

if tilting > 8500 and tilting < 10000 then
High B4
pause 1

else
Low B4

endif

if tilting > 2000 and tilting < 3500 then
High B3
pause 1

else
Low B3

endif

if tilting > 350 and tilting < 400 then
High B2
pause 1

else
Low B2

endif

'++ Repeats the main loop ++
goto main
```

Phone Cord

Cut one end off the phone handset cord and strip back the jacketing so that about 2 inches worth of wire is exposed. There should be four wires. Insert these in a pairs of two through the extra mounting holes in the tilt switch. The two sets should break down as follows:
Set 1: Green and Yellow
Set 2: Red and Brown

Next you are going to want to strip a little off the end of each wire and solder them into place. Use this guide:

- Green connects to the black wire from the switch
- Yellow connects to the four resistors
- Red to connects the long leg of the phototransistor
- Brown connects to the shorter leg

When they are all soldered into place, insulate the connections and set things in place using epoxy. Make sure to pay extra attention to the spot where the phone cord connects to the base sensor as this joint will receive a lot of strain when you use it. Use a tab of extra epoxy to glue the phototransistor into place on the top of the sensor.

Set the sensor aside to dry, it's done for the time being. It's now time to work on the strap.

Start making the strap

The strap holds the sensor in place on the user. It's adjustable using a Velcro closure and is made of soft neoprene so that it's flexible and comfortable.

Cut a strip of neoprene 2.5″ side and plenty long (more than one foot). Using a zig zag stitch to allow for stretching, fold the neoprene in half and sew a seam along the edge. You should now have a double thick strip of neoprene that is 1.5″ wide.

Trim any excess fabric from the waste edge of the seam.

Using your finger, invert the strip so that the seam is now on the inside and you're left with a nice neoprene tube.

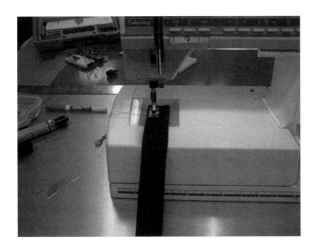

Then, sew the two ends of the 3/4″ elastic in place on the strap so that it holds the tilt sensor tightly in place, forming a 1″ loop.

We basically made a small cradle or harness for the sensor out of elastic.

The strap is now complete.

Sew Velcro

Sew a 1″ - 1.5″ piece of 3/4″ wide Velcro onto one end of the strap and the corresponding piece of velcro onto the other. Adjust placement for optimal fit.

Trim off excess neoprene from the strap.

Build your relay board

The relay board receives the signals from the sensor and triggers the now hacked Atari controller.

Make a board to house the 5 relays using this .cdr file. To make this board, use either wood, acrylic or standard circuit board material and drill out the holes as pictured.

Insert the 5 relays into the center of the board. The 5 relays are controlling each of the four cardinal directions and the fire button.

Don't connect Pin 1 to the PIC chip yet. Instead just attach a red wire to Pin 1 on each relay.

Also, don't connect the Atari controller cable yet (as pictured).

Make elastic loops

Sew one end of the 1/4″ wide elastic banding onto the strap in the center. Cut the elastic at about 1.5″ in length and sew it onto another short piece of the 3/4″ wide elastic.

Wire together Pin 8 on all of the relays using black wire.

Wire together every Pin 7 with black wire. Extend a long black wire from Pin 7 on Relay 1.

The board has some extra holes to thread the wires through. Using these holes will prevent connections from breaking.

(Please not that the relay really only has 8 pins in total. So, when I say Pin 8 I am really referring to what should be Pin 5. I'm not exactly sure why they give you 3 invisible pins on each side, but that's just how it goes I guess.)

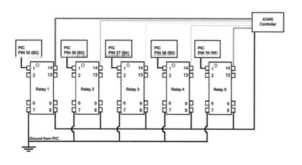

Open the controller

Open the Atari controller.

Remove the circuit board.

With diagonal cutters or a pair of pliers, remove any excess plastic that isn't used to fasten the controller shut.

Jack it

Cut a small square hole into which you will insert the telephone handset jack.

Glue this in place with epoxy and/or hot glue.

Hack the Atari controller

If you are lucky, your Atari controller will be labeled nicely like ours was to inform you which wire connects to which button.

If it's not labeled, it is fair to assume that the black wire is ground and the 5 remaining wires are for the four cardinal directions and the trigger button.

You can figure this out easily by trial and error. Start a game, plug in your controller and connect each wire, one by one, to the black ground wire and see what happens on the screen. You should find that each wire represents a different direction (or the trigger button). If it doesn't, you have selected the wrong ground wire and you need to keep trying.

Controller to relays

Wire the Atari controller to the relays.

Pin 8 on all of the relays should be connected together, and in turn connected to the black wire from the controller. This is because all of the button controls on the controller board share a common ground pin.

Pin 14 on each relay should have a different colored wire from the controller soldered to it. In other

words, each relay should be connected to a unique colored wire.

(Please note that the relay really only has 8 pins in total. So, when I say Pin 8 I am really referring to what should be Pin 5. I'm not exactly sure why they give you 3 invisible pins on each side, but that's just how it goes I guess.)

Build a circuit board

Build a circuit board for your PIC micro controller using the two schematics listed below.

The 20 MHZ component is a resonator and not a crystal. It should be as close to the chip as possible. To make it work with a PCB you may need to bend the middle ground leg forward and the other two in towards each other so they can fit into the socket side by side.

All of the relay connections on the PIC diagram connect to Pin 1 on the relay board diagram. For instance, Pin 35 connects to Relay 1, Pin 1.

The rest should be pretty straight forward.

Be careful not to miss any wiring the first time around (like I did).

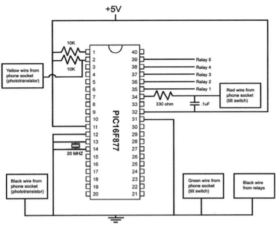

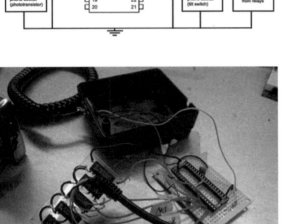

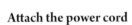

Program the chip

Program your PIC chip with the code you tested earlier.

Attach the power cord

Cut the end of your power transformer.

Locate the positive and ground wire inside the transformer's cable and expose a little bit of wire from each cable.

Plug in the transformer and test to make certain you know which wire is which and that the voltage written on the transformer is correct (sometimes they mislabel it).
Once you are certain that the polarity and voltage are corrected, solder it to the board.

Transfer the chip

Transfer the chip from your development board to your circuit board.

Be careful not to bend any pins!

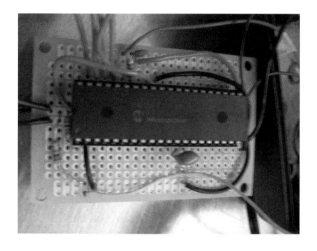

Troubleshoot

Test to make sure it works.

If it works continue on.

If it does not work check the following:
- Is it plugged in?
- Are your relays working? If you listen carefully you should hear them click?
- Is your chip programmed?
- Is your wiring correct?
- Are there any crossed wires on your board? Is you chip getting hot?
- Is your phone socket wired correctly?
- Did you leave off any components?

Insert tissue

Remove the 'stick' from the top of the old joystick box and replace it with some tissues. This is for form and function.

Close the case

Using hot glue, glue the red button in place permanently - this is for form only.

Cut a small notch in one side for the power cable to pass through.

Cover the PIC chip in electrical tape to prevent it being shorted by the relay board.

Knot the power cable and insert the circuit board and relay board into the case.

Put the tissue-enhance top on and screw it shut.

Coat the sensor in silicone

The base sensor needs to be protected from certain environmental conditions, so we decided to coat the whole thing in silicone. This also helps protect all of the fragile wires and wiring.

The Art Stuff brand silicone we used gets mixed in a one part to ten ratio. I made a small batch of 20 grams of part A to 2 grams of part B. I used a really small amount of blue silicone pigment to turn the silicone from clear to blue. Note, the dye is for form only.

Mix the two parts of the silicone together thoroughly and paint it onto the base in a thin coat. Apply it all over the sensor; accept for the top of the phototransistor.

Use multiple coats to achieve proper protection. Allow it to dry overnight between coats.
Remember to wear nitrile gloves and to work in a well ventilated area.

Putting it all together

Insert the tilt mechanism into the loop on the strap.
Plug the phone cord into the jack in the Atari controller.
Plug in the power brick and attach the Atari controller to the console.
Attach the neoprene strap and sensor to the user.
Insert the battery into the emitter ring and place it on your ring finger.
You are now ready to begin playing.

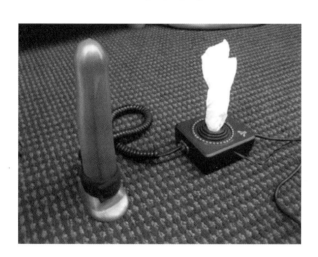

Play

Play Atari with yourself and friends in the manner that you always dreamed of and never have to decide between sexual stimulation and video games again.

Allen Stein

MORE'S LAW

I was in my office in Seattle's oldest business district, one that had served men since the 1850's and the Alaska gold rush. It harbored every manner of bar, saloon and tavern for wild women and wilder nights. Then the 1990's brought the Web gold rush and the area became an Internet startup haven – historic with easy-to-find bars – just what tech geeks wanted. It was laced with T1 and some of the best tech infrastructure of Seattle.

I had been inventing. What devices could I possibly control via Internet commands? A sex machine?

A year went by of tech revs and testing and happy friends, and then word spread about my work. I remember the moment. The latest Playtime magazine came in my mail and an awesome cover headline struck me – the cover feature was about us, right below a sexy picture of porn star Nikita Kash. So cool. I began to read aloud.

"Allen Stein has probably given more women more orgasms than even the most prolific porn star- male or female. And he's done it all without ever having had sex with them. What's his secret?"

My secret? A little project called thethrillhammer – one of the first commercial Internet controlled sex machines online and the first sex machine to be a feature performer in pornography. I had assembled a collective of Pacific Northwest designers, artists, and sexual technologists to build what would be the first of many cutting-edge pleasure craft under the brand name thethrillhammer. And it was all being created from this historic office that had seen over a hundred years of sensual satisfactions. Now we were spreading sexual satisfactions to the world. And we were the cover story. I found myself a big drink easily.

The pathways to innovation often begin in strange places.

Historic gold-rush Seattle is one beginning for myself. But earlier I was already being shaped by growing up in Des Moines Iowa... apparently guiding my life toward orgasms. These are some of the stories and history. Sit back.

Fade in:

Arse Elektronika Sunday Oct 4th 2009 6 PM
Noisebridge Hacker Space San Francisco, CA

Noisebridge is up a flight of stairs or an old freight elevator. People are working on various types of projects throughout the vast space. Some have computers in front of them. Others have all sorts of electronics. Solder smell wafts through the air and the sound of keyboards click away as the drone of many conversations fills the air.

In the middle of the space Kyle Machulis finishes up an impromptu rambling on sex toys. He shows us how to fuck a building using GIS technologies. He also shows off what he thinks is a never been used next generation virtual sex device for men called the Real Touch. Little does he know? He is telling the crowd he wants to hack the device and make it a 'car wash' for his Hot Wheels collection. Allen rips the Real Touch out of his hands and refuses to give it back.

The always affable Johannes makes a brief introduction of sensual technologist Allen Stein and he takes the stage.

Allen Stein: How did I come to be asked here to talk about what I do? Band camp. Yes my friends, band camp. I was a teenager at the dawn of the information age, hitting puberty full swing. I put away the Pong and I began to using ANY available technology for the purpose of exploring sexual pleasure.

My main method of courtship was the telephone. This was the first technology of seduction for teenagers all around the world. The hazards when I was a kid were hairy palms, nowadays its blisters on your thumbs.

It must have been 6th or 7th grade when I got access to computers. My first was an Osbourne at my friend Kirk's house... you know the one with the tiny 5" green screen and flashing cursor. Next was an Apple][and a][+ that my dad used for his pan pizza restaurant, running Visi-Calc, one of the first spreadsheet programs.

I soon was a fixture in their office as I loaded the computer with Ultima by Lord British and refused to budge for hours.

Then we got a computer at home. Color monitor! Two drives! Able to "back up" my programs to keep a copy at my friends' house. And the cherry – a 300 baud modem and CompuServe account. Could I ask for more? "Want to play a game?"

I started exploring BBSs and learned very fast that I wasn't the only one using technology for pleasure. I remember seeing the first ASCII porn. "If I squint I think that's a nipple, or is it a hyphen?"

My journey continued into the wonderful world of sex technology in high school. I discovered the bass trombone and the tuba. Most other high school band players, especially the trumpets, were extending their range higher and higher. I did the opposite…went low. I filled in the bottom with my deeper rumble and was very amused that my genitals would tingle on some lower notes. The lower I went the better it felt. I particularly noticed it with the low 'D' below just the staff. Wow! I was playing a giant musical vibrator! This feels good.

Why low 'D'? Lots of music that you feel in the your crotch is in 'D.' In fact many rock guitar and bass players use a Drop-D tuning to get a sound that can rock you like a hurricane. (Drop D, is an alternate guitar tuning style in which the lowest string is tuned down or "dropped" one whole step to D rather than E. It is used widely in modern rock and heavy metal but turns up in many other types of music.)

Perhaps the body's energy systems respond to different vibrations. The sacral chakra located in the sacrum is associated with the genitals and sexual energy. This energy system vibrates at 'D.' Coincidence?

Then my testicular tingling started to make sense to me. Fairly basic stuff perhaps. Sound is a travelling wave that's an oscillation of pressure, vibrations. My testicular tingle? Resonance.

Beyond the tactile buzzing, you get these moments of pure bliss when you play for a big audience. I hit the international stage in high school at the Montreux Jazz festival playing in a big band with a group of some of the most amazing musicians I have had the privilege of playing with. There is a moment when the band is on and the crowd is digging the scene and you as a musician get this rush that continues well after the end of your set. This awesome buzz sometimes takes a couple of hours to wear off. Just so happens to be the same cocktail you get from sex. Adrenaline, Dopamine, Vasopressin.

I think the sex and music connection is further explained by the guitar player solo fuck face. When guitar players are playing their solos they make their fuck face. Check it out next time you see a show. Fuck faces.

I think this is when these initial seeds in my brain were planted for what was to come.

I was a heavy magazine geek in 1993. Mondo 2000, Future Sex, and an old Meckler Media magazine – VR World. That's where I saw the first ad for some kind of sex simulator driven by a 486 computer, selling for $25,000. Cool.

It was either CompuServe or The Well were where I started to notice the work of Norwegian media artist Stahl Stenslie who many consider the 'father of cybersex.' Stahl built amazing cyber-suits that delivered a variety of sensations, called cyberSM, connecting participants in Paris and Cologne in a world's first "multisensory, full-body communication system." He used graphical representation to create an interface to control the sensations, most likely over 9600 baud modems.

Now fast forward a few years later to the tech boom of the 1990s. I had an enjoyable career in high tech. Worked for at the time a half billion dollar global direct marketer of hardware, software, and peripherals and then another reseller who had to rebuild his business after Apple changed the way they sold their hardware. Remember the Apple clones? I ended up at the home automation, remote control and wireless camera giant X10. You old timers should remember our 'little' innovation of the 'pop under' campaign with the scantily clad young lasses and 'Put Anywhere' wireless cameras. I swear I had nothing to do with it. Soon thereafter I was, as they say, dot bombed. Thus began my brief period of chronic entrepreneurship.

It was during this period I went on my annual Frank Zappa kick. I would start listening to his work for weeks on end. One of my all time favorite Zappa works is his opera *Joe's Garage*. If you are not familiar with this work I will give you a quick synopsis as I am sure it was partially responsible for cementing the vision of sex machines into the dark recesses of my brain.

This is the short version with no spoilers.

It's about what would happen if music were to become illegal.
The main character Joe meets this girl who dumps him.
He meets this other girl name Lucile that gives him an unpronounceable disease.
He claims he got it from the toilet seat.
He gets depressed and turns to religion for help.
He goes to L Ron Hoover at The First Church of Appliantology.
Hoover determines Joe is a "latent appliance fetishist." Joe asks if he should "come out of the closet."
He is instructed to "go into the closet" to achieve "sexual gratification through the use of machines." Joe finds a sex machine he likes from Germany (because all the really good ones come from over there). The sex machine is named Sy Borg, a XQJ-37 Nuclear-Powered Pansexual Roto-Plooker.

XQJ-37 Sex Machine. Oh I've got to get one of them. Then one evening I was watching Woody Allen's movie *Sleeper*. The movie shows an 'orgasmatron,' actually a snazzy elevator. Once entered, it contains some future technology that rapidly induces orgasms. They needed this machine because almost all the people in the Sleeper universe have become impotent or un-orgasmic. (According to Woody, in the year of 2173 the males of Italian descent are considered the *least* impotent.)
That night of web surfing revealed early internet-controlled vibrators on a website called ifriends.com. By using suction cups little photocells could attach to a computer monitor to sense the color of blocks on the screen. When they changed color from black to white it controlled the speed of little sex toys.

Then it happened. I had my epiphany. I had to tell my wife.

This is how she said it happened. "My husband said I have another million dollar idea. I rolled my eyes. Then I was floored by his answer. He wanted to build the ultimate vibrator, one that really works. No faking it. He wanted to remote control this sex machine via the internet. He wanted to mount this thing in this antique gynecological chair. Oh no not this gyno chair again. I thought I was finished with that idea. Besides it sounded like something out of Barbarella. Explain please? He wanted to build it into an antique gyno chair that he tried to talk me into buying for him many years prior for Chanukah for his yacht. Don't ask. His idea was to attach a very strong vibrator to the chair and allow end-users to control the pleasure of others who rode it. He was going to charge a per-minute fee for the service. He must have had this planned for months because he had major details planned. He assured me he had done his market research.

"I had my reservations. So I placed a wager, 'Allen, if you don't pay more than $50 for that gyno chair, write a business plan, and find an investor, then you have my full blessing.' The very next day he found the chair he wanted in a Seattle tattoo shop and they were ecstatic to unload it for fifty bucks. Apparently it didn't work like they had hoped for tattooing. Imagine that. He promised them passwords to the site. It was a 350 lb beast that had to be brought home. Then he stayed up all night writing and plopped a business plan in my lap. He also found an investor. I lost that wager but he found a passion."

That I did. We incorporated and rented a loft in Seattle's artsy post industrial Georgetown area. Then it came together pretty quickly.

This first machine was purpose built to be a self-contained commercial sex device. We planned it for a studio where clients could take turns pleasuring the mostly female riders. We upholstered the seats with thick cushions and sparkly fabric to make it as comfortable as possible for our talent to work shifts of chatting and fucking online. We designed HR "Gigeresce lights" to illuminate the scene and provide another level of interactivity. We used X10 Internet controls to turn the lights on and off. I happened to have more than a few modules sitting around the house. We mounted a flat screen to a long twisted metal pole attached to the bottom of the chair so it would be at eye level to the rider. On top of that pole we mounted a high-end pan tilt zoom video conferencing camera so the viewer and the rider could have eye contact. The positioning allowed the viewer to see everything – they could zoom in on a nipple or inspect the quality of a Brazilian wax.

The initial video conferencing platform we used was Emulive which worked until they changed their licensing policies. So we eventually developed APIs to integrate into commercial adult cam systems. We went through a handful of providers over the years, cams.com, camz.com, flirt4free.com.

I used a Sybian as the heart of the gratification device. If I was going to build a better vibrator I better start with great components. We mounted it on its side in a stainless steel case and started building the internet controls.

We tried to use WebPower's ifriends.com approach and hooked the Sybian to light sensors. It worked, but not very well. Screen bleed. The coolest part of that concept was the ability to play a sex machine like a Theremin.

With our first trade show quickly approaching and budgets running thin I needed a better solution fast. We decided to use the serial port. The parallel port was on its way out; USB was early in its spec and on its way in. We hacked directly into the motor controller boards and fried the first of a couple Sybians.

We decided to go analog. We built a knob turning robot and wrote a script to control it over the web. The machine was able to produce mind blowing orgasms, controlled from anywhere in the world. We launched the site and ushered in a new era of commercial teledildonics.

"Teledildonics" refers to technology-aided sex. The term has become a catchall term for anything from virtual reality suits to remote-controlled vibrators. There are telemetry devices and synchronization devices. Frankly being called a teledildonist is just plain tacky. Can't we come up with a better word for it? Tele means from afar. I get that. Good prefix. Cyber I like. Cyber-kubernetes "steersman," perhaps based on 1830s French cybernétique "the art of governing" is good too. Dildonics. Dildonics. Dildonics. What the hell people? Dildonics. I don't think it is even a word. I'll check.

I just checked a couple resources for us. I got a "Did you mean Delmonico?" from dictionary.com. Why yes I would love one. Medium Rare. I checked the Oxford dictionaries next.

"Sorry, there are no results for that search."

This is our chance to correct the wrong set so long ago by Ted Nelson. He can give that word up. He has given birth to many others and doesn't need this one anymore. Hypertext, hypermedia, virtuality,

Intertwingularity are all great words. Teledildonics. Not so much. It is just plain tacky.

Here are some suggestions. Phalanetics. Nope just as bad. Biocybernetics. Perhaps. Cybernetics. Like it. Better. My personal favorite is "Cyberotics."

What is happening in 2009 in the field of cyberotics?

We have seen slight innovations in cyberotics although most of the commercial players have shuttered or moved on. Sinulate's site is up but their service is down. Butter Butter from Brooklyn dissolved amid launch. A new video conferencing player called DoLips launched over a year ago integrated with HighJoy devices. Their site is still up but they have no performers online at the time of this writing. No performers, no traffic, no money.

The secret to monetizing Internet-controlled sex devices in these business models lies in the proliferation of devices to active performers on video chat sites with heavy traffic. I am happy to report a year later from the last Arse the live chat segment of the adult entertainment industry is still doing better than ever with even without cyberotics.

Trends? We are seeing sex toys that use some basic adaptive learning. There are many new vibrators that remember your sessions for playback. Experience based entertainment. The Real Touch device is a synchronization device that ties sensations in the genitals to the action on your screen. It was designed and built not to exploit the technology but to solve a huge problem. The Real Touch a direct result of people accessing stolen content (porn) on tube sites and peer to peer networks. The adult industry is losing money as more and more people jerk off for free.

It seems people feel *entitled* to their pornography much like feel *entitled* to their music. There is a whole new generation that have always thought that music and porn was free. What changed the business models of the music industry is now happening in the adult industry. Business models are adapting and changing.

The future lays in experience or educational based entertainment rather than plain passive forms. The timeline of innovation is compressing, making all this development happen faster. Think about it, it's only four to five generations since B&W TV in 1950 to Multichannel Multimedia and Personal Computing. It was close to 23 generations from the first factory printed book to radios, telephones, record players, and movies. Development is far faster as the media delivery systems evolve so profoundly – from cable TV delivery to CD-ROM, to Video CD, to DVD, to Blu-ray, to web, and today to mobile that are being replaced by smart phones every 18 months by consumers. Moore's law. I've seen cybersex move from BBSs to Newsgroups to Websites to NetMeeting to Webcasts to Peer-to-Peer to Game Sex to thethrillhammer and to modern cyberotics. I call that More's law. Yes please, I would like some More.

Sex technology is about to take giant leaps forward. The market is ready. Sensuality and sex have become mainstream media topics – on Oprah and The Today Show, as well as cover stories on *US News & World Report* and AARP magazines. We are seeing more explicit content is more socially acceptable. Sexperts are emerging as mainstream media stars. What Dr. Ruth started is now a rich field of sex and relationship stars like Dr Laura Berman, Dr Pepper Schwartz and Violet Blue.

The adult products industry is performing well in the recession, according to *Forbes* magazine. As consumers look for ways to slash spending, many are heading to the bedroom to keep themselves entertained. Sex toy sales continue to grow in most countries.

These indicators lead me to believe we are about the ride the perfect wave. The timing is right to find a new 'sticky app' for cyberotics. I believe that future will be found in devices that are personalized to each user, supporting them with an infrastructure to encourage them to live richer, fuller, and more sensual lives long into our old age.

My prospering from the delivery of the ultimate orgasm in adult entertainment chat systems has come to a close. A new chapter has opened.

Reflecting back on my career I definitely may be a contender as to delivering the most orgasms than the most prolific porn stars male or female. At this point 10 years later I must be well beyond both porn stars, rock stars, and athletes combined.

I think it is time for some quid pro quo.

Ping me.

Tatiana Bazzichelli

ON HACKTIVIST PORNOGRAPHY AND NETWORKED PORN

Hacktivism is a term derived from the union of two concepts: hacking and activism. Hacking is a creative practice; an irreverent and playful way of using computers which might also address an ethical and cooperative modality of relating to knowledge; activism indicates individual or collective action for achieving social goals and developing political battles. Not only does hacktivism address a certain use of technology, but it might also be seen as an attitude in which exchange, sharing, freedom of information and experimentation are central elements.

If we for example consider the hacker and activist backgrounds of the Italian and Spanish underground culture of the past thirty years, the 'hacktivist' attitude is very often connected with the radical-punk idea of self-management, DIY and independent production. Many of the hacker and activist utopias spread in 1980s and in 1990s, took shape through collective experiences in social centres, squats and virtual spaces such as BBSes (Bulletin Board systems), and promoted the idea of building up self-organized and economically independent spaces of networking and interactions [1].

The idea of creating nets of relations among individuals and collective experiences where radical use of technology is connected to radical politics is not just limited to the creative use of computers and technology. Sex might also be seen as a working field of hacker experimentation and a context in which to express the DIY punk approach. While hacktivism is the direct political and social action online, pornography becomes the direct political and social action on one's own body (outside and within the network).

Some experiences in the European queer and activist culture (mainly in Italy and Spain, if we consider their diffusion on a national scale) showed how to transfer this experimental hacker and DIY attitude from technology to the body and to the broader concept of sexuality. Experiences where the DIY-structure of the punk scene, and the hacker ideas of sharing, openness, decentralization, free access to information, and the hands-on imperative (Levy, 1984) became a challenge to create a different kind of pornography [2].

Punk Porn as an Activist and Art Strategy

Pornography is the obvious demonstration of how different dynamics of sexual power are written and perpetuated through the body. But it can also become a territory of both radical and playful action, as some of the experiences in performance art, the post porn scene and the later queer culture demonstrate. A thread that connects practices and theories, from Genesis P. Orridge and Cosey Fanny Tutti to Richard Kern and Lydia Lunch, to Annie Sprinkle, Bruce LaBruce, Maria Beatty, Emilie Jouvet and Beatriz Preciado, just to mention a few [3].

Contrary to the 'traditional' approach of a type of feminism which considers itself politically correct, but which perpetuates and consolidates a traditional dichotomy between genders and doesn't resolve the problem of power structures - pointing out porn as a form of exploitation - it is necessary to start to

---

1 For a more complete history of the hacker underground communities in Italy, see my book: (2006) Networking. The Net as Artwork, Aarhus, DK, Digital Aesthetics Research Centre, 2009. Online at: www.networkingart.eu/english.html. Some of the reflections in this text are developments of the topics discussed in the last chapter of the book ('Extra Gender', translated from Italian to English by Maria Anna Calamia).

2 Some Italian examples in the past ten years: the cyber-punk-transgender Helena Velena (www.helenavelena.com); the activity of the women and gender laboratory Sexyshock in Bologna (www.ecn.org/sexyshock / www.betty-books.com); the queer-party experience of Phag-Off and what followed it (www.myspace.com/warbear); Tek-festival (www.tekfestival.it); the Pornflakes collective (www.pornflakes.org); Carni Scelte (www.myspace.com/carniscelte); Vida Loca Records (www.myspace.com/vidalocarecords); and, experiences in Berlin which follow the same Italian thread: Poopsyclub (http://www.myspace.com/poopsyclub) and Sabot°Age (http://www.sabotage-berlin.com), just to mention a few. In Spain: Girlswholikeporno (http://girlswholikeporno.com); Ex-dones (www.exdones.blogspot.com); Post Op (http://post-porno.blogspot.com); GoFist Foundation (http://gofistfoundation.pimienta.org); Diana Pornoterrorista (http://pornoterrorismo.blogspot.com); La Quimera Rosa (http://laquimerarosa.blogspot.com).

3 To know more about the above-mentioned practices and theories, see the list of books and essays suggested in the bibliography.

consider pornography no longer as an enemy, but as something which should be appropriated from below.

As long as we continue to label pornography as sexist, chauvinist and macho, we will be leaving it in the hands of those who really want it as such, making it a mirror for relationships of hierarchical power between men and women, and a territory of repetitive banalities.

to personally enter inside the mechanisms of bodily expression and the production of desire, to subvert them from within, in favour of a fluid dimension; a mirror of a fanciful expressive everyday life.

This concept of porn is not a recent invention, but follows in a long tradition on the fringes of alt porn and amateur porn, and has been taking shape through the net (blogs, p2p technologies, platforms of video

CUM2CUT 2006 snapshot

CUM2CUT 2007 snapshot

As Stewart Home pointed out, quoting Cosey Fanny Tutti:

> 'You get feminists saying you're being exploited and all the rest of it. But it's not like that. It's a total power trip. When you are being exploited, it's when you are doing something you're not comfortable with. Where it's not you. Where someone is saying 'do this'[4].

The answer, and the challenge, might be in making porn punk, or rather making punk porn: Pornography becomes an experimental field open to all of us, and an occasion for breaking the dynamics of crystallized power, self-governing one's own sexuality. The conscious role of women (and men) in this vision is

and photo sharing, etc.) and social networks since the middle 2000s. As Katrien Jacobs wrote in 2007, in the introduction of her book on netporn:

> 'Porn is successfully being appropriated and reinterpreted by alternative producers and active sex workers, young pro-porn feminists, queer porn networks, aesthetic-technological vanguards, p2p traders, radical sex/perv culture, and free speech activists'[5].

But describing the process of bringing porn studies into the humanities, approaching the field as participant observers, and making it the subject of 'alternative porn' or 'post porn' events, Florian Cramer argued:

4 Stewart Home: Confusion Incorporated. A Collection of Lies, Hoaxes & Hidden Truths. Hove, UK: Codex Books, 1999, p.70.

5 Katrien Jacobs: Netporn. DIY web culture and sexual politics. Lanham, Maryland, USA: Rowman & Littlefield Publishers, 2007, p.3.

'The price for such integration is the avoidance of all conflict. Whether as a provocation, as an expression of the power of sex or of sexual politics _ what is thus liquidated, the obscene, was what marked the points of intersection between the experimental arts and commercial pornography, in Courbet and Duchamp, in Bataille's novels, Hans Bellmer's dolls, Viennese Actionism, Carolee Schneemann's *Meat Joy*, but

zone' of individuals who are not very excited by the often male-oriented mainstream porn, neither are they part of the queer / alt porn communities.

The DIY-punk approach of the post porn, netporn festivals and queer communities often resolve itself in creating a new aesthetics, with the result of making porn more accessible, but to a narrow group of people. The challenge is in which way it will be possible to

CUM2CUT 2007 snapshot

CUM2CUT 2006 snapshot

also in pornographers later honored as artists, such as photographers Nobuyoshi Araki and Irving Klaw, fetish comic strip artist Eric Stanton and sexploitation moviemakers Russ Meyer, Doris Wishman, Jean Rollin (whose work was honored by Aida Ruilova during the most recent Berlin Biennial) and Jess Franco[6].

To my view, this position is true – pornography is still expressing itself through a dichotomy. It is the mirror of the desires and the needs of two very different categories of people: Those who like commercial pornography or those who like 'alternative' pornography. It doesn't touch a larger group of people, the 'middle

'open porn to everyone', rather than making it a field of study among specialists – or a successful niche market within the porn business.

One of the strengths of punk culture was to open the concept of art to all; everyone could sing or be a pop star. You didn't need to be pushed by the musical industry, at least in the beginning of the phenomenon. This is the punk approach we should still follow. This is where the vision of Hacktivist Porn comes from.

Hacktivist porn might seen as an opportunity to invent the 'porn of the future', reconstructing its meaning beyond gender stereotypes or specific political and sexual inclinations. It might be the challenge to make porn an everyday practice of life, which belong to all, like Fluxus artists did with art in the Sixties and Seventies.

6   Florian Cramer: Sodom Blogging: Alternative Porn and Aesthetic Sensibility. In the C'Lick Me Reader, Amsterdam, 2007, which can be downloaded at: http://www.networkcultures.org/clickme/pdf/clickmeReader_9MB.pdf.

In 1972 Wolf Vostell, one of the pioneers of video art, happenings and the Fluxus movement, wrote in a postcard: 'Duchamp has qualified the object into art. I have qualified life into art'. We should qualify porn into life.

Networked Porn: The experience of CUM2CUT – Indie-Porn Short Film Festival

The concepts of openness and Do-It-Yourself were the starting points for the development of punk culture and hacker ethics, but also for networked art such as mail art, for example. The art of networking was based on the figure of the artist as networker: a creator of sharing platforms and of contexts for connecting and exchanging. It was not based on objects, nor solely on digital or analogical instruments, but on the relationships and processes in progress between individuals. Individuals who could in turn create other contexts of sharing. The same Do It Yourself hands-on practice was used to describe subsequent phenomena of networking and hacktivism; from Neoism to Plagiarism, up until the 1990s when the network dynamics were affirmed on a broader level through computers and the Internet.

Inspired by the 'hacktivist attitude', which referred to an acknowledgement of the net as a political space, with the possibility of decentralized, autonomous and grassroots participation, Gaia Novati and I created the experience of CUM2CUT: Indie-Porn Short Film Festival (indie as in 'independent'), where we wanted to apply the hacktivist attitude and the idea of open network to pornography.

From www.cum2cut.net, 2006:

'CUM2CUT is an independent pornography competition, a four day marathon in which participants are invited to make a short film. The short film must be pornographic, but participants don't need to be actors or porn experts.

The main idea of CUM2CUT is to connect a heterogeneous and international network of people who wants to create art independently and to freely express their sexuality. Therefore, CUM2CUT is a networking platform, which aims to create open and free artistic networks where people can address the topic of pornography without being marginalized.

CUM2CUT is an opportunity for participants to play with sexuality and pornography by producing and enjoying indie-porn-short-films: the best short-films you'd want to make and to watch! An expert jury formed by people involved in porn/queer subculture, visual art and experimental cinema select three winners, whom will receive a prize.[7]

CUM2CUT started in the city of Berlin, in October 2006 as part of the Porn Film Festival Berlin, a week-long event managed by the German filmmaker Jürgen Brüning.

To ensure that the films were made during these four days, the participants had to follow a few rules – incorporating specific symbols, objects, sentences and sounds, and had to randomly pick a 'genre' for their films: Superhero XXX, Splatter Porn, Pop Star Porn, Alien Porn, Horror College Porn, Dadaist Porn, Christian Porn, Socialist Porn, 9/11 Porn, and so on…

The jury, formed by people like Joe Gallant, Julia Ostertag, Shu Lea Cheang, Ela Troyano, Tessa Hughes Freeland, Shu Lea Cheang, Francesco 'Warbear' Macarone Palmieri, and many others, saw the films for the first time at the premiere like everyone else and selected the winners also considering the audience reactions.

In November 2008, the CUM2CUT films were part of the exhibition 'Porno 2.0', at the D21 Art Gallery

7   CUM2CUT: Indie-Porn Short Film Festival, 2006, http://www.cum2cut.net/06/index.html.

in Leipzig, and in 2009, CUM2CUT started to travel around Europe as a nomadic entity. The CUM2CUT films have been shown in many international festivals and conferences (see http://www.cum2cut.net/en/index.php for a complete list of the screenings), and last year (2009), we were partners of the Arse Elektronika Festival and the films were selected for the Prixxx Arse Elektronika at the Roxie Theater in San Francisco.

this context, queer means to express sexuality beyond boundaries of identities and to cross the limits of fixed genders and stereotype.

At the same time, the idea of queer is closely connected to D.I.Y. punk and hacker culture: CUM2CUT wants to encourage everyone to express themselves using their bodies and media from an independent point of view, creating new

CUM2CUT 2006 snapshot

CUM2CUT 2007 snapshot

What is the concept behind CUM2CUT and why do we connect it with the above-mentioned idea of hacktivist porn? This is how we presented CUM2CUT in 2007, when we organized two different porn marathons, one in the city of Berlin (the porn competition) and one on the net (the pr0n competition):

> 'Unlike mainstream porn, the event CUM2CUT in Berlin focuses on the activities of the international independent and countercultural queer movement and presents a platform where artists, filmmakers, DJs, actors and everyone interested can collaborate.

> This happens through the activities of people who are part of the queer countercultural movement. The goal of the festival is to develop and enjoy new forms of subversive body-politics in terms of gender and sexual orientation. In

experimental queer languages. The idea is to broaden possible imaginaries and 'desirescapes' related to pornography.

The concept of play is directly linked to CUM2CUT: body-games and sex toys are the starting point for subverting the mainstream expression of pornography, through the practice of détournement. CUM2CUT aims to reach those who are fed up with mainstream pornography, who are frustrated by tolerating the rules of normalized society. At the same time, CUM2CUT provides a critical perspective on political imagination and rejects the notion of radical politics necessarily being boring and serious.

CUM2CUT proposes an experimental concept starting from bodies and space, spreading pornography as clouds of pollen to eroticize the city of Berlin, mixing up fluid bodies, nomad identities and playful sexualities. This year, CUM2CUT also takes on the challenge to eroticize the network, through a Pr0n competition, with the aim to sharing strategies, shortcuts, tricks and pranks, to 'dress up' the technology, turning it into a porn tool. Our bodies are not the only interfaces for expressing sexuality and desires, therefore why not try using other tools? Let's think about technology itself as a porn subject!

The Indie-Porn Short Film Festival is open to queer, homo, hetero, lesbo, bisexual and transsexual independent video- and filmmakers, pornographers, performers, screenwriters, directors, queer cultural activists and artists, hackers, free thinkers, excited minds and all the people that like to mix technology and bodies. The purpose is to develop ideas through film production and to share them with an interested community as a participatory exchange. Thus, in order to open not only the boundaries of sexuality but also those of artistic expression, the videos produced must be licensed under the Creative Commons.

Much like how the idea of peer-to-peer is expressed by the P2P symbol, we have created the term C2C (cum to cut). C2C closely links pleasure and orgasm to the action of sharing pornography experiences'[8].

CUM2CUT Indie-Porn-Short-Movies Festival

CUM2CUT press logo

Since we decided to release the films under Creative Commons, they are all available online at: www.cum2cut.net/en/index.php?sect=movies2007; www.cum2cut.net/en/index.php?sect=movies2006.

CUM2CUT was not only an occasion to rethink a different kind of porn, but also a collective workshop where the participants in the competition, the members of the jury and us – the porn-networkers – where joined together in collaboration.

The end result was a kaleidoscopic assemblage of pornography, which most of the time hardly could have been considered porn in its traditional connotation.

CUM2CUT was a challenge to try to see porn as an open concept, as a new way of living the city space creating a network of people through pornography, and an occasion to disrupt the boundaries and the limits of sexuality. It was an attempt to open the

CUM2CUT flyer 2007

8   CUM2CUT: Indie-Porn Short Film Festival, 2007, http://www.cum2cut.net/en/index.php?sect=background.

concept of porn to everyday life and to all possible personal interpretations and derivations.

Towards a Hacktivist Pornography

Trough the idea of hacktivist porn we demand a different vision of sexuality, creating projects, products and creative actions for which pornography becomes an artistic platform, an autonomous and experimental network. Pornography is the tool for exchanging and sharing sexual experiences.

Pornography, eroticism and sexual pleasure become opportunities in which to play, into which to bring a form of open desire, expressions of pleasure of the different men and women, or different expressions of sexualities not invented yet, in which diversity and otherness can cohabitate. To interact with pornography together with many other open identities, which go beyond the sterile definitions and stereotyping of gender, means to hack porn from within to make it a new form of free artistic experimentation. It is an expressive occasion in which people can consciously choose their own role.

This vision of porn, Hacktivist Porn, invest all kind of sexual preferences - heterosexual, bisexual, homosexual, transsexual, and all kinds of porn; indie porn, queer porn, netporn, bear porn, dyke porn, kinky porn, science fiction porn, dada porn – what you think porn is – and all kinds of sex-conscious and sex-positive forms of desire, opening up to as many people as possible.

Hacktivist porn means to try to imagine the porn of the future, in which everyone has the right to express her or his own selves. It doesn't avoid conflicts, but it shows that it is possible to create a personal vision of porn formed by polyphonic desires.

Bibliography:

Beatriz Preciado: Kontrasexuelles Manifest, Berlin, Germany, b_books, 2003.

Florian Cramer: Sodom Blogging: Alternative Porn and Aesthetic Sensibility. In the C'Lick Me Reader, Amsterdam, 2007.

Katrien Jacobs: Free Passwords: The Bumpy Guide to Porn Sharing, in Neural Magazine, Bari, Italy, 2009. Online at: www.libidot.org/neural/passwords.htm.

Katrien Jacobs: Netporn. DIY Web Culture and Sexual Politics. Lanham, Maryland, USA: Rowman & Littlefield Publishers, 2007

Stewart Home: Confusion Incorporated. A Collection of Lies, Hoaxes & Hidden Truths. Hove, UK: Codex Books, 1999.

Tatiana Bazzichrelli: Networking. The Net as Artwork, Aarhus, DK, Digital Aesthetics Research Centre, 2009. Online at: http://networkingart.eu/the-book.

Tim Stüttgen (Ed.): PostPornPolitics: Queer Feminist Perspectives on the Politics of Porn Performance and Sex Work as Cultural Production, symposium reader, Berlin: b_books, 2009.

Kim De Vries and Pepper Mint

POLYAMORY + TECHNOLOGY = SEXUAL AND ROMANTIC ABUNDANCE

Discussions of sex and technology often center on innovations in sex toys, or special effects in porn, but technology plays a far greater role than that. Technology increasingly facilitates social connections in general, and romantic or sexual connections in particular. A group that arguably has been among the most enabled by technology are those practicing polyamory. This paper begins to explore how various technologies enhance and empower polyamorous relationships, and considers how polyamory may be in turn shaped by the use of technology. Among the platforms considered are online calendars, social networking sites, chat clients, microblogs and blogs, and also less obvious technologies like matching algorithms for speed-dating, dating websites, and Massively Multi-player Online Games (MMOs).

The approach combines both theory and lived experience, as the speakers adopt the double vision of both participant and observer in the world of polyamory. Many of the issues considered may arise in monogamous relationships as well, but are more prevalent in polyamorous arrangements, or may require novel approaches when multiple relationships are involved. Unsurprisingly, the practice of maintaining more than one emotional and sexual relationship at a time can lead to greater complexities in scheduling, to maintaining long-distance relationships, and often may leave individuals feeling isolated in locales that don't have much of a poly population nearby (Rust, 486). Technology has been used by the poly community in numerous ways to address these issues.

The internet has been used as a device to collect disparate non-monogamous folks into a cohesive community.

In the last few years polyamory has become far more visible thanks to widespread online coverage in podcasts, internet TV series, and informational websites. This exposure has helped many people not living in the most accepting or informed locales still learn about polyamory and realize that they are not alone. "Oh wait there's a word for that? That's funny,

I've been doing it without a word for five years." Many may be more comfortable discussing a possible poly identity online than in person, from coming out as poly to introducing the idea of group sex, kinky practices, or attending play parties. In fact one of the primary uses of poly sites might be to help people talk about poly sex, because as has been widely observed, people are often more comfortable talking about intimate matters online. Eventually some will feel ready to publically call themselves poly, and a growing number of social networking sites accommodate this.

Poly Weekly

Even Facebook now offers the option of defining yourself as being in an open relationship, while other less mainstream sites, such as Openly or Tribe.net have long welcomed poly-identified members and event listings. Because all of these sights also attract people based on many other identity markers, the concept of polyamory is introduced in a relatively non-threatening, "by the way" manner. The spread of information in this indirect way may lead to increased acceptance of polyamory by those who might resist a more direct introduction.

The ongoing spread of polyamory and creation of poly community might not have happened without the internet.

Online forums tend to act as poly advice forums. Most active online poly discussion seems to be non-monogamy advice. Because there are so few agreed upon models of poly relationships, the poly community is being formed specifically to help people to figure out what kinds of poly relationships are workable. These sites are crucial to the formation of a poly identity because they provide the same kind of blueprints usually supplied by cultural images for more common relationship types.

People stumble across online poly communities (usually, an online advice community) and that helps them realize that they might formalize and/or start talking about their existing non-monogamy, which tends to lead them to identify with poly. For example, many podcasts that address polyamory and discussion forums as well center around participants calling or writing in with a description of their situation and requests for validation and/or advice. Validation of a poly identity or relationship is often lacking for poly people living in regions where the poly population is small and finding this validation online enables those trying to live polyamorously to better resist entrenched cultural patterns.

Because there are so many people in homegrown non-monogamous arrangements, with no words or associated identity, poly websites are attractive as a first step in developing an identity that includes non-monogamous behavior. In the last few years, several books on polyamory have been published, each with a website that links to other resources (Block, Taormino). As those authors have toured and done interviews, they have drawn even more attention to polyamory. Even negative coverage of polyamory publicizes its existence.

Technology and internet serve as poly relationship facilitating mechanisms.

Along with its benefits, polyamory brings increased complexity and time-management challenges. Many poly groups have found online match-making or dating sites, online calendars, instant messengers, chat clients, and VOIP to ease some of the difficulties one may experience while managing multiple relationships. Every relationship needs attention in order to be sustained, but poly relationships may require more attention because they fall outside recognised models, and so may not receive the social support that comes with such recognition. In particular, when people are known to be part of a couple, that relationship is socially supported.

For the polyamorous, relationships that are not publicly revealed may be interpreted as cheating, while people who are openly poly may fine that those around them interpret this as meaning there are no boundaries around the relationships. Further, while members of a couple will be expected to attend each other's family functions, work-related parties and so on, members of a larger polyamorous arrangement do not receive the same recognition. It they are to be included, some explanation is almost always required, and possibly extensive negotiation. This lack of public recognition may then cause poly relationships to need even more direct reinforcement from each partner. In this sense they exhibit a dynamic parallel to that of online relationships.

Jonathan Marshall has argued that participants in online communities and relationships often experience "absence" or ontological uncertainty experienced online because "there is no marker of existence beyond the act of communication itself (Marshall 2004)." The key similarity is the lack of a marker. However, online communication and other online applications actually offer ways for poly people to mark their relationships.

For example, some poly constellations share an online calendar in order to coordinate who is seeing whom on which days each week. Not only does this help to ensure that scheduling conflicts are minimized, and that fairness (however that group defines it) is observed, it also is a visible sign that everyone with access to the calendar is included and recognized as part of the group. Similarly, marking multiple connections on a social network, through the relationship labels available, or through the use of status updates, wall posts, and so on. In the screen shot below, we see that one of the co-authors has labeled his relationship status as "in an open relationship with" his partner.

View Photos of Pepper (67)
View You and Pepper
Send Pepper a Message
Poke Pepper

Information

Relationship Status:
In an Open Relationship with

Facebook clip

However, while Facebook offers "open relationship" as a category, it does not allow you to actually list more than one partner, thus still shutting out the other partners in a way that might actually feel more excluding than if the open status could not be marked at all.

But while ensuring fairness might be an issue, and marking a relationship's existence is important, perhaps the greatest challenge related to time is simply finding enough time to give multiple relationships the attention needed to sustain them. If all people involved are working full-time jobs, there may be at most five weekday evenings and two full weekend days which can be devoted to interacting in person. If people have children or any other obligations, this time may be even less. Online interactions through email, chat, or playing online games together can mitigate this challenge to some degree, by creating a feeling of being together and sharing experiences even when not physically together.

With those who are physically proximal, we can easily exchange hugs or handshakes, share meals, go to museums or engage in hundreds of other physical activities which because they are public and common may not seem terribly significant or intimate. But shared physical experiences of any kind cement bonds between people, and also reveal a great deal about the participants to each other. We have an ongoing feeling of being together, or 'co-presence" (Zhao 2008). When we cannot be physically together

with a loved one, whether because of time constraints or distance, various internet technologies can help assuage the need for contact.

Unsurprisingly, video chat can come very close to meeting in person, allowing you to see facial expressions, body language, and so on, and any communication that occurs via a chat client, whether text, voice, or video, can be recorded, leaving a reassuring sign of the contact behind. In many ways the exchanges on Facebook seem to also stand in for physical encounters – going to lunch or for drinks, attending cultural events, etc (De Vries). But neither of these avenues of exchange captures the feeling of sharing an experience, of doing something together. Interestingly, it is becoming clear that playing computer games is not very different at all from actually experiencing those events. Research on the therapeutic use of computer games reveals that users very quick come to see and avatar as an extension of their own body, and the experiences users have in a game affect them as much as those they have in real life (Yee and Bailenson). This suggests that people who play together in online games will feel as though they are really experiencing a shared activity in addition to any communication that takes place during that time. Certainly my own experience in online games supports this view. The use of both computer-mediated communication and people in relationships has been investigated by a number of scholars, but the use of MMOs as part of relationship maintenance has just begun to receive scrutiny.

My own observation and anecdotal evidence suggests there is considerable overlap between the Poly community and those playing MMOs, and for years the connection between MMOs and poly issues has been reported (anon., 2001, Torrone, 2004). For example, in the screen capture below, we see and excerpt from a page devoted to the marriage of three

I have been married to Shamhat and Coeur de Leon for quite some time now. They are both a lot of fun, although Shamhat seems to be very engrossed in her library studies as of late. I have known them both almost from the first instant I set foot on this island, long long before the war.

Clan Lord screen capture

characters in Clan Lord, one of the first MMOs for the Mac beginning in late 1996 and persisting to the present. Along with actually enacting poly relationships in a virtual space where they might receive the social support lacking elsewhere, players may make use of mods that allow them to mark their status. A recent addition to World of Warcraft mods is one offered that allows players to list multiple relationships in their stats.

Future Work

Because technology has figured so heavily in the development of polyamory, those who are comfortable with technology seem significantly represented in the poly community. The use of technological affordances to support polyamory is an area ripe for further research. We might also question whether being a technophile depends on some set of traits that also inclines one toward polyamory, and how the reliance on technology might be shaping the community. A case study might also be productively conducted to explore how internet technologies are bringing together marginalized, disparate, and/or stigmatized groups.

Internet technologies bring things out into the light that people might not be willing to discuss in person. Sometimes it seems safer, but also because in some settings, such as MMOs, participants may build trust through shared experiences. The internet's "will to speech" can aid those who are not yet comfortable speaking in person about these issues to still participate in the community, and more importantly to communicate about polyamory with their loved ones.

Even on platforms that are not explicitly devoted to poly issues, if the technology in some way leads to people feeling connected, building trust and forming relationships, it may create circumstances in which people feel comfortable disclosing a poly identity, and perhaps testing it in the relative safety of a virtual space. At the moment, many online spaces seem safe and welcoming, but will that safety persist?

Polyamory may be more accepted precisely because in the popular media, the sexual aspect is downplayed while the multiple-love aspect is foregrounded. However, as polyamory becomes more visible, the sex-positive attitudes on which the community is founded and the potential for sexual abundance afforded to members of the community, will certainly draw fire. Will technology continue to support acceptance, or will it be linked to a more negative vision of poly as a threat to traditional values, just as technology is often derided as threatening traditional social relationships? Even in that case, technology and polyamory seem likely to remain firmly entwined.

Works Cited:

anon. "Confessions of a Virtual Warrior" *(Details)* A Journal of Art Criticism, Published by the South Bay Area Women's Caucus for Art Volume 7, Number 2, Spring 2001.

Block, Jenny. Love, Sex and Life in an Open Marriage. Berkeley, CA: Seal Press, 2009.

Adam Briggle, "Love on the internet: a framework for understanding Eros online", Journal of Information, Communication and Ethics in Society, Vol. 6 Iss: 3 2008.

De Vries, Kim. "Your Friend has just tackled you. Bite, lick, or tackle them back, or click here to theorize about what this all means." In *datadirt: media, culture, technology* http://blog.datadirt.net/2008-11/guest-post-by-kim-de-vries-your-friend-has-just-tackled-you/

Hales, Kayla D., "Information and Communication Technologies and You: Multimedia Relationship Maintenance" . AMCIS 2009 Doctoral Consortium, Paper 22. http://aisel.aisnet.org/amcis2009_dc/22

Hian, L. B., Chuan, S. L., Trevor, T. M. K. and Detenber, B. H., Getting to Know You: Exploring the Development of Relational Intimacy in Computer-mediated

Communication. *Journal of Computer-Mediated Communication*, 9: 00, 2004.

Kathleen E. Hull. *Same-Sex Marriage: The Cultural Politics of Love and Law*.
Cambridge University Press, 2006.

Marshall, Jonathan. "The Online Body Breaks Out? Asence, Ghosts, Cyborgs, Gender, Polarity and Politics." Fibreculture Issue 3, 2004. http://journal.fibreculture.org/issue3/issue3_marshall.html

Rust, Paula C.. "Monogamy and Polyamory: relationship Issues for Bisexuals," in *Psychological Perspectives on Lesbian, Gay, and Bisexual Experiences* (2nd ed.). Edited by Linda D. Garnets and Douglas C. Kimmel. New York: Columbia University Press, 2003.

Taormino, Tristan. *Opening Up: Creating and Sustaining Open Relationships.* Berkely, CA: Cleis Press, 2008.

Torrone, Phillip. "Love and romance in the world of Massive Multi Online Role Playing Games." *Joystiq*, Oct. 6, 2004.
http://www.joystiq.com/2004/10/06/love-and-romance-in-the-world-of-massive-multi-online-role/

Walther, J. B. . "Computer-mediated communication: Impersonal, interpersonal and hyperpersonal interaction." *Communication Research*, 23(1), 3-43, 1996.

Yee, Nick, and Jeremy N. Bailenson, "The Difference Between Being and Seeing: The Relative Contribution of Self-Perception and Priming to Behavioral Changes via Digital Self-Representation." *Media Psychology*, 12:195–209, 2009.

Zhao, Shanyang & Elesh, D. "Copresence as 'Being With': Social Contact in Online Public Domains." Information, Communication & Society, V. 11, No. 4 June 2008, pp 565-583.

Websites Consulted:

http://azriel.puddleby.info/marriage.html
http://www.deltatao.com/clanlord/
http://www.facebook.com
http://www.fileplanet.com/209136/200000/fileinfo/Dragon-Age-:-Origins---Polyamory-Mod
http://www.openly.us
http://tribe.net

Robert Glashuettner

VIENNA RUBBER CITY

Austria's capital not just offers architectural marvels, delicious domestic food and overall a high quality of life but also a high density of top-notch latex fetish fashion designers.

Tackling the difficult question about the true psychological origins of fetishism, it is fitting that rubber devoted Vienna has bred the foundations of psychoanalysis through Freud over a hundred years ago. Not being particularly interested in the roots of their own enthusiasm, though, the protagonists of the Viennese latex fashion community cater more for their own unique designs and workmanship. Throughout the last 20 years, there has been a notable rise of latex manufacturing in Vienna that has established a strong impact within the fetish community on an international level. Thanks to the famous Life Ball, the HIV-fighting and glamorous fetish and queer event founded in 1992, the Austrian rubber community once every year also has a big public outlet for their self-made products. But it is not just about fashion fantasies that you like to share with friends and acquaintances on special occasions. The successful labels *HW Design*, *Simon O.* and *Studio Gum* also willingly care for those special needs, the realisation of kinky and sexually charged latex gear. It is the mixture of handmade, elaborately designed pieces and their extreme and bizarre siblings that maintain the lustful tension between fashion and sex in a very tempting way.

Studio Gum

Talking about latex fashion production in Vienna, you have to start with Johannes Möller. The former second hand dealer and carpenter founded his heavy rubber Studio Gum in 1990, after experimenting with the liquid raw material for some time before. Today, the now 60-year-old Möller maintains a niche within the latex community that is still internationally unmatched, due to his craftsmanship and provision of extreme fetishistic desires with thick and heavy rubber hoods, suits and bondage accessories. Johannes Möller, together with his long-term employee Mike, does everything himself from scratch, thus being both a good source of quality rubber sheeting as well as serving as an inspiration for younger latex designers. Meeting him in his small factory in Vienna's second district, Möller works away while answering various questions.

Mr. Möller, you are the godfather of the Viennese latex manufacturing community.

Yes! I was the first one who founded his company, in 1990.

Why did you found Studio Gum?

I tried to buy myself some latex clothing but couldn't find what I was looking for. Then I stumbled upon liquid raw material and started to experiment with it. I had put together some pieces quite quickly and then tried to sell it. Soon I thought: Why not turn this hobby into a job?

Was it a hard decision to openly show your fetishistic interest through Studio Gum?

It wasn't easy at first, but my kids were already grown up and my wife supported me. Also, the fetish community really started to take off at that time, so it actually came natural in a way.

Did you have to answer a lot of questions throughout those first years? I figure that especially people around your neighborhood were wondering what exactly you were doing.

Of course. But in the end, there were never any problems. The vice squad did check up on me, though, because I placed advertisements in special interest papers and magazines. I could have been a brothel in disguise, they figured. So they came and checked everything and that was that.

What happens throughout a working day at Studio Gum?

My day starts at about 5am when I check my inbox and answer e-mails. This way, I can already communicate with the Japanese and can still communicate with the Americans. Afterwards, I am here in the studio by about half past seven.

You have a lot of international customers.

I couldn't even pay my rent with the customers from Austria. My customers come from all over the world. I am kind of a market leader with all the heavy rubber stuff that I'm doing. Nobody is as specialised as I am. That is also why I supply to so many international wholesalers.

You have a long-term employee and your son also works here.

Yes. I am doing the suits, Mike takes care of the masks and my son prepares the raw material and the individual parts we need here in the factory to manufacture the items and pieces.

Heavy rubber is a special form of latex fetishism that is less known in the broader public and not as established as the more mainstream side to it which is derived from fashion. Do you feel a stigmatization?

Yes, I do. It is easier with all the fashion stuff to get some positive recognition. It is perceived as a fashion thing and looks good and everything. But the heavy stuff with masks and gas masks is in close relationship with the BDSM community. Then, for some weird people, you aren't far away from being a child molester. So, you have to be cautious. But I live with it. I don't have to tell everybody, anyway.

Talking about this aspect, has anything changed throughout the last 20 years of Studio Gum?

It's not as hard anymore. In the beginning, sometimes people were really uptight to buy stuff. But with the rise of the web it is now much more convenient than back when I still had to have my printed catalogues.

Do you ever change your product line?

Barely. I have the same range of goods for years now. Other things just didn't sell well. I chose to perfect the pieces I am famous for and now do everything myself, from scratch.

You also have a huge collection of gas masks available for viewing at Gasmasklexicon.com. Why don't you sell those masks at Studio Gum anymore?

Because we don't have any supplies left. The big collapse was with 9/11 where all the old gas masks that were stocked in Europe were sold to the US. Apparently one wholesaler even sold 40000 Israeli civilian respirators. On September 15th of 2001, all gas masks were gone. As for the gas mask lexicon and the supply I had before:  People always asked me if I'm crazy because of such a big supply. In my best times, I piled up about 500, 600 respirators, to sell and utilize them for my other latex pieces and masks. There is a rule I preserved from my times as a second hand dealer: Some things you just don't get immediately when you need them. So you have to have a proper supply depot.

Where do you get the liquid latex from?

It gets delivered in barrels. I am buying the liquid latex through a company in Vienna which gets it from another company in Germany. Originally, it comes on a tanker from Malaysia.

On your website StudioGum.com, you promise that you always try to fulfil the rubber fantasies of your customers. But how do they develop confidence in you? This is a very personal thing, after all.

With the possible anonymity when communicating through the internet, this is no problem at all. They just write to me what they want. It is rarely the case that anyone really comes up to my factory. Sometimes, though, international visitors from Japan or the US top their trips through Europe off with a visit of my studio. But when they come here, they are mostly disappointed, because there is nothing to see, really. This isn't a showroom. We are a manufacturer and not a physical shop.

Of course. And sometimes also funny things happen. Once a female customer called me and said that at the very moment, she was playing with her husband and wanted to strangle him but it wouldn't work. He could still breathe and was already starting to giggle and laugh. She had mixed up the valves on the rebreathing bag. After my advice, she arranged it in the correct order and said: "Ah! Now he's shaking! Thank you!" – and hung up the phone. But it can also be annoying. Once a customer was complaining about a suit I made for him. It turned out that he was wearing it backwards! I just made him a new one and sold the other one as used. This way, nobody could say that I provide bad service.

Is heavy rubber mostly a male hobby?

Yes, I would say 98 percent of my customers are male.

What do they write when they order something online?

It's just normal orders. Just with a few people I am wondering if they should really use my stuff. Some might just hurt themselves. You have to have at least some kind of affinity to it, otherwise it gets complicated. You have to know how to use all the tubes and valves.

Do you advise your customers how to properly use your pieces?

Where is the fascination of latex and heavy rubber derived from? What is your personal experience? Is it the bizarre design, the body manipulation or a mixture of both?

It is hard to separate. Talking about myself, I can only suggest to people that if you are a latex enthusiast, you shouldn't pursue this professionally, as a job. Because when you arrive at this point where you can have all your fantasies and dreams fulfilled and realised, it just wears off. It did work well for a number of years. But then for me, at some point it made a "click" and it was like a switch turned off.

So, you turned off your own enthusiasm yourself, unintentionally?

So to say, yes. I developed this perfectionism: Everything had to be exactly the way as the fantasy in my head told me to. And when it just didn't work that way, I lost interest. When you have this idea, for example: I want to be enclosed in latex for many hours, not being able to move, and then you realise that it is physically not possible, and then you get disappointed. This is how it was with me, anyway. And then you say: If this doesn't work, I am not interested anymore.

What about the others, the typical Studio Gum customers?

It is very diverse. There are simple people all the way through to academics. Nevertheless, most of my customers have enough money; otherwise they couldn't afford the expensive latex pieces. One guy once told me that he put his normal suit over his latex suit and you couldn't tell that he was wearing rubber underneath. That was part of his fetish.

What do you do after you retire in a few years?

I'd like to start some other business. At the moment, I have lots of fun repairing old pocket watches, especially those which are heavily broken. I really enjoy the moment when I am successful at bringing an old watch to life again.

HW Design

After some years of collaborating with Studio Gum, HW Design founder and former commercial artist Harald Wilfer for once decided to take on a latex fashion company of his own. Having fond memories of his tailoring grandmother, creating heavy rubber and latex fashion pieces with standard cuts was just the very first step for him. Hiring alumni from the Hetzendorf fashion school in Vienna, HW Design quickly became well known for their sophisticated, complex and edgy cuts that are rare in the world of latex clothing. Oscillating between kinky and classy outfits, the broad collection from 57-year-old Wilfer, his wife Sandra and their team is a very special feast for the eyes and of course a pleasure to touch. Meeting the man in his spacious office and factory in Vienna's green, recreational area of Kaisermühlen and around us the female employees were busy at work during the interview.

Mr. Wilfer, why did you start out with HW Design?

It was around 1998. I pitched fashion designs to the German latex fetish magazine Marquis. Afterwards, Studio Gum would always put together the pieces. At some point, I decided to start my own company as Johannes Möller from Studio Gum wasn't interested in the fashion side of rubber clothing anyway. But he provided me with the material and taught me the craft of how to put together latex fashion pieces. Also, because Marquis was producing own films, they needed a place to shoot which we could provide. So, one thing led to another.

Was it hard to quit your former day job and fully concentrate on your own company?

Doing the latex stuff was fun. Not so being a commercial artist for 25 years. Many things changed in terms of computers and software tools. More and more young people took on professional assignments to receive very little pay in return. For this amount of money, I wouldn't even bother to turn on my computer. So I decided to go for the things I really enjoyed – I could have always gone back to my old job. But it worked out very well and it wasn't long after I was able to hire my first employees. This was the kick-off of HW Design.

How did you come up with the idea of designing latex pieces, anyway?

Fashion is rooted within my family. My grandmother had a tailoring with 100 employees before World War II. My sister and my mother also learned the tailoring skills from her. I grew up surrounded by sewing machines. It didn't interest me in my younger days but still, fashion design always came kind of natural to me. I know how something is supposed to look. That is why I decided to just hire employees who have an education in fashion. As for the fetish side of it: Well, what is fetish? It could be cars, shoes, leather – almost anything. We do latex clothing.

Do you want the public to perceive fetish fashion in a new way?

Of course. If you look at it, 20 to 30 years ago, if a woman wore thigh-high leather boots, almost everybody would say that she was a prostitute. Today, this is part of fashion and completely established. I think it will be the same with latex. Some years ago, there was this idea of the pervy latex fetishist who wears his latex masks in front of his TV alone, drinking beer, and wouldn't dare to go outside. Today, in almost every issue of Vogue, you can find at least one latex piece, whether it is gloves, stockings, leggings, whatever. It becomes part of everyday life and fashion.

How do you come up with a design of a new piece?

Sometimes my wife, my staff or I, myself, come up with ideas; sometimes customers request special custom designs. It is often the case that a female customer who wants a fancy dress for a special occasion will visit us to discuss the design. This is also the reason why I need experienced, talented staff because my own skills do not suffice for these precise design decisions.

It would seem that latex clothing is ready to fully merge with the fashion world. Still, your designs are mostly geared towards the fetish community. The before mentioned single latex pieces in fashion magazines do happen, but it seems if there is no real mutual interest and communication happening. Does this bother you?

I don't know. It's true that the majority of our customers are fetishists who love the material. And I honestly can't think of, for example, a society lady who decides to wear latex then quickly starts sweating and wants to leave it on. She would probably rip off the clothes, shouting: "I won't wear this crap anymore!" On the other hand, female pop singers and performers such as Lady Gaga have a different use for it; they use it for their stage shows and videos. As for latex clothing and the fashion world: Some companies tried it once but it was a disaster. They had wonderful, colorful latex pieces – but together with buckles made out of metal, the oxidation process started. Then you get blotches on the latex you won't be able to remove.

Apart from your sophisticated fashion designs you also have a lot of kinky and bizarre latex pieces.

Yes, of course! We need to make some money, too.

So, the kinky stuff is the real cash-cow?

Yes, people still spend the most money for the naughty stuff. And it's a good thing when you deal with customers who have enough money in the first place.

Do you also meet your customers in the office or abroad?

Sure, mostly when we attend fetish events like the German Fetish Ball, Rubber Ball or Fetish Evolution. But you tend to see the same people. It is almost like some kind of a travelling circus. It's a big family, a community.

Do you enjoy being part of this community?

Yes! Kumi, our model from San Francisco who visits Vienna every few months, also frequently tells us what the people like at the moment. My wife Sandra, who is also well known in the fetish community, also travels a lot. There is a lot of contact going on between the models, specifically.

What is the relationship to other, international latex manufacturers?

We have good connections and there is a real friendly competition going on. We are friends with one of the biggest manufacturers, Demask from the Netherlands, because we played with open cards from day one, so there is a mutual respect going on. When we're in London, we visit Libidex, House of Harlot and other designer houses we like and respect. Everybody makes their own unique clothing range. Of course there are black sheep but they mostly come out of China.

There are latex designers in China?

For the past one or two years, they are copying our designs like crazy. The quality is bad, but they do try. And they also use our images; they just steal them from the online shop.

Is Vienna an important location for the international latex fashion community?

No. The most important location is and will remain London. London is kind of the fetish headquarters. Everything comes together there, whether it is shoes, latex, leather or whatever.

For almost a decade now, your wife Sandra and you also have your own erotic pay site Fetish-Live.com.

We started off at a time when there were basically no pay sites in Austria. It was a pioneering work what we did back then. We also made videos for some time, so we had the latex fashion designs, the erotic website

and the film production. The movies spawned even more interest in our clothing range. All in all it was a big success story. But eventually, the pirated copies of our videos were spread around the net and brought our movie production to a hold. Because VHS was dead, the DVD popped up online just a few weeks after we started shipping them. Then you start asking yourself if this still makes any sense. We shot our last film in 2007 but we continue to maintain our website Fetish-Live.com. It still has a good advertising aspect to it. But we couldn't live on this alone anymore.

Discontinuing video production must have been a little bit sad for you as it sure was a personal thing, too.

Of course. You don't do these things just to get rich; we've always had lots of fun. It would be false-faced to say that this was and is just a professional endeavour. It also wouldn't work that way. We've seen a lot of websites come and go but we've been online for a decade now.

Does HW Design have more male or female customers?

As for the non-kinky fashion pieces, we have more female customers at the moment. The reason for this is that our design range is more focused on women's clothing. Of course, the transgender- and transvestite-communities are also into our designs because they often like to be pretty and style themselves extensively. During the Life Ball in Vienna, we have lots of customers, both male and female.

Do you sometimes also get orders from film companies who need special outfits for their shootings?

Yes, we once had a lot of pieces shipped for the third Matrix movie. But they just bought stuff off our online shop; they obviously didn't need any custom made pieces.

Has HW Design been listed in the credits?

Unfortunately, no. This would have been great, of course. But this isn't an easy task to achieve;

sometimes the people on these projects also tend to forget those things.

Coming back to the tension between fashion and sexuality once more. Do your customers openly speak about their latex fashion and fetish interest?

Not really. On one hand, it is and should remain something special; on the other hand, people shouldn't be ashamed. Some people are really frightened!

Why?

Because of the public's opinion. They fear that someone could see or catch them while they are wearing latex jeans or something similar. This is something we do want to change. There is really nothing weird about people wearing neatly designed latex clothing.

How does it affect you when you are in contact with latex every day? It must have been or still is your personal passion, too. Has something changed over the years?

Of course. The kicks I had 10 or 15 years ago are almost gone. You tend to get a professional view on things. When I see a woman in a latex outfit today, the first thing I ask myself is: "Where did she get her pieces from?" It takes a longer time nowadays to really realise just how great something looks. There is no denying that a part of my sexual fantasies have been taken away from me. I turned my fantasies into reality, so it is not fantasy anymore.

Is this something you suffer from?

It is nothing that doesn't affect you and doesn't make you think, of course. On the other hand, I am still having lots of fun doing my job and I can't imagine that this will change in the future.

Simon O.

Being the youngest among the Austrian latex fashion manufacturing companies, 40-year-old Simon Pöltl and his wife Manuela should not be mistaken as newcomers to the scene. Their history with the attractive and well known brand Simon O., derived from the founder's former surname before the marriage, dates back to the late 1990s. The couple recently moved from their long-term street shop in Vienna's 17th district into an idyllic house in lower Austria's small village of Weyerburg. Being self-taught and having a good sense and talent for fashion design, Manuela and Simon are specialised in affordable, clean and well-cut basics made out of both thin and thick latex sheeting in many different colors. Having the manufacturing facilities and the office separated, the interview with Simon takes places in a room so tidy you'd think a photo shoot would have had taken place recently. And there is a good chance that it really has.

Mr. Pöltl, what was the starting point for Simon O.?

A friend and I were looking for latex fashion and, of course, I first stumbled upon Studio Gum where I also bought the first latex dress for my former girlfriend. Then I thought: This is glued together and I am quite dexterous – why not try it myself? Coming up with a good cut is not easy at first but something like a miniskirt shouldn't be that much of a problem. We found a place in the UK where we could order latex sheeting. We bought some of those and started gluing latex together in my living room. At first we did easy things like basic tops, pants and leggings for private use and to give to friends. Soon my friend was like: Why don't you sell this stuff? At first I was appalled by this idea, but then, it just progressed and at one point the living room was no room to live in anymore.

This was in 1997?

Yes. And then a sex shop owner took notice and told us he wanted a collection. He had latex fashion before but the brand had gone out of business and now he wanted something new. We already started experimenting with colors at that time. Red and black was common, but other colors like yellow, blue and purple were rare. So the shop owner was quite excited and asked what we could offer him. We didn't have a company back then. So we came up with a collection and printed it out. There were shirts, pants and other basic pieces, but we still didn't have any unique cuts. So we came to his place with some single pieces and said: "Well, we have this and that in different sizes." - But really, all of that didn't exist at the time!

So you actually made it up.

Yes. But of course, then we really had to do it. The shop owner ordered basically everything we offered him in different sizes, for two stores. We tried to keep cool and said: "No problem." But we needed money for the material and so we said to him that we would need an advance of 50 percent of the final amount. This was our seed money! And, of course, after this incident we had lots of work to do.

What did you do before, where did you get the crafting skills from?

I did completely different things before. I finished an agricultural school and received education in car plumbing afterwards. Then I designed kitchens and worked as a facility manager. I had lots of spare time doing this job. This led to the latex experiments.

Was it easy to decide whether or not to dedicate your life to the design of latex clothing?

You need to have that special interest, of course. You don't just start out doing latex fashion without personal background behind it. Upon shipping our first collection to the sex shop, we became a company. Then there was this problem of in which industry they would file us. We didn't fit into the tire industry and we also did not qualify for tailoring because we don't sew anything. So they came up with a new industry name, very complicated. "Production of latex clothing", but with a lot of exceptions so that we would not interfere with other industries.

You seem to have a good balance of aesthetic latex fashion and more fetishistic, bizarre pieces that you both sell via your website Latex.at.

You could say that. We have a mixed range of customers. Some of them prefer the fashionable pieces because it just looks good. Then you have the other half who is very much into the material. Apparently, there are more women in the first group and men in the second.

How did you manage to establish the brand Simon O. so successfully?

I don't know. Really, it's hard to explain. Of course, we always had good quality products; this is one of our most important features. Then we were always well known for having a wide range of different colors. Today, this is nothing special anymore, but it was when we started out. The German fetish magazine Marquis featured us in an article and this also helped us rise in interest within the fetish commu- nity. And from there on, it was just a natural progres- sion. But it surely  wouldn't have worked if we had only been after money. Nowadays, we can live on it, but it's been a long time coming.

In many cases, heterosexual men want to introduce their fetishes to their female partners.

Well, there are of course also women who are inter- ested without someone leading them to latex fashion.

Because latex is glossy and looks sexy, the fashion side of latex is generally very attractive to a lot of women. Really wearing latex, on the other hand, is something different, because you have to know how to correctly put it on so that it's not exhausting and feels uncomfortable while you are wearing it. And, yes, many heterosexual men who are interested in the material want their female partners to wear latex clothing and try to introduce it to them step by step.

With fetish clothing, there is always this strong tension between fashion and sexuality. Some customers might be shy and frightened when they attempt to buy latex pieces. How do you take away their shyness? Or isn't this that much of a problem?

It's not that bad. And it depends on who you're talking to. Sometimes it's the other way round. There are people who are very open about their sexual interests and desires. So, when someone asks for some hot pants with a built in anal condom, you should remain

But there is this moment.

Yes, definitely. And, as I said, some people are absolutely uninhibited and you have to deal with that, too. Some guy, businessman-type, came into our shop one day, wearing a suit, and asking for a latex diaper. We don't have that in our collection but we said: Yes, we can make you one. But the conversation didn't end there because then he desperately had to tell us why he needed the diaper. It turned out that he frequently attends board meetings and when they have to vote on something or whatever he likes to completely let it go. This is how he gets his kicks. In situations like this you're supposed to be like: Well, it's his thing and as long as he doesn't hurt anyone while doing this – why not?

There is little communication and exchange between the usual fashion world and the fetish fashion community. What is your take on that?

calm and act as if it is nothing special at all. When you ask: "Is it for fisting? With one or two hands?", you instantly feel the relief of the person. Then the chat becomes more relaxed.

The main problem is that latex can't be industrially processed. If a fashion designer would try to do a latex collection, it just would be immensely expensive. Everything has to be hand-made. Also, you need to know how to handle the material correctly and take care of it so that it lasts long without getting worn out. But famous designers do use some latex basics here and there, mostly for their fashion shows although sometimes it's just used as an eye-catcher.

How many employees do you have?

All in all, we are six people. Manuela is the boss and takes care of accounting, communication, orders and everything else that takes place in the background. And she is our model, of course. Design details are being discussed together. Sometimes our customers demand some special features that might also be good ideas for future designs. Our four employees then produce the pieces.

What education have your employees received? What knowledge and skills do you demand from them?

Primarily, they have to be dexterous and use their heads while producing the pieces so that everything works out properly. You need to do it for a while and get experienced. Latex can be hard to process, especially if the material is very thin. When a new employee starts out, we let him or her practise with leftovers first. One of the first exercises is to just glue a straight line, later you try round lines.

Manuela and you also maintain your own erotic pay site RubberVita.com. How did this project start out?

It started while making photos for our webshop. We thought: Why not do a real photo shoot? And then we just did and it developed over time. Soon our subscribers came up with their own ideas such as wearing tight jeans over latex stockings which we also provided most of the time. It's fun and it has a good PR effect although it is additional work for us, of course.

You rarely work with external models on RubberVita. com and feature mostly yourselves in the often quite explicit photo sets and videos. How do you separate this from your private sex life?

At first, we just started doing this and hadn't talked about whether there would be an interference with our personal love life or not. Thinking about that now, I think it is not that easy to separate. Because you don't do this with a stranger on a professional basis but with your own partner, and it's mostly things that we also like to do privately. And, of course, you have a commitment to your subscribers. So, when there just isn't that much time or one of us is sick, there certainly is pressure to say "Hey, now we really need to do a new photo gallery!"

Johannes Möller from Studio Gum suggested not turning one's fetishistic desires into reality because the excitement starts to wear of over time.

I don't know if this will happen to me, too. Of course, you tend to develop a different view on things. That's not a bad thing as long as your partner still excites you and this is definitely the case with me.

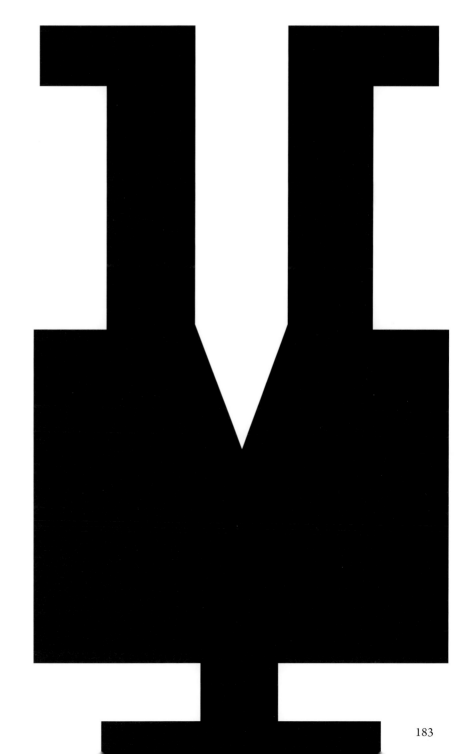

Jack Sargeant

FIFTY YEARS OF NAKED LUNCH

"It was a hectic, portentous time in Paris, in 1959, at the Beat Hotel, No. 9 rue Git-le-Coeur. We all thought we were interplanetary agents involved in a deadly struggle… battles… codes… ambushes."

- W S Burroughs, *The Western Lands.*

The following is drawn from a series of typed and handwritten notes for a public lecture given at the Mu-Meson Archives, Sydney, to celebrate the fiftieth year since the publication of William Burroughs' novel Naked Lunch.

Neither simple biography nor mere hagiography this paper is an exercise in psycho-cartography, in creating a map around the radical Burroughs that authored *Naked Lunch*. This is a map that follows lines of flight and uneven trajectories, a map with no centre but endlessly fluxes and flows, a map that picks up speed but doesn't have a border.

In April 1959 the first seven astronauts were selected for the emergent US manned spaceflight program. In July, only two months after this announcement, a book was published in Paris that would change literature: *The Naked Lunch.*

Jack Kerouac's *On The Road* was published two years previously and, by 1959, the cliché of the finger clicking, goateed, bongo playing, reefer smoking, sandals and shades wearing beatnik was well established. Articles on hipsters and beatniks appeared in various publications, from Yvette Vickers, the Beat Playmate photographed by Russ Meyer for the July 1959 *Playboy,* to regular articles in men's magazine and jokes in *Mad,* through to more considered reviews of works in newspapers. But, in 1959, William Burroughs, the elusive figure who already had something of a cult reputation following his 'appearance' in Kerouac's *On The Road* and a dedication in Allen Ginsberg's *Howl And Other Poems* (published in October 1956) had vanished.

William Burroughs had left the United States for Mexico in 1949, before heading far deeper into uncharted territory both psychological and geographical travelling into South America in 1951 searching for the psychotropic vine Yage, to North Africa in 1954 and finally to Europe in 1958. His belongings could fit into a single case should he need to evacuate a residence at speed, bag in one hand, typewriter clutched in the other.

Numerous aspects associated with Burroughs can already be observed at play: space travel, beat culture, pulp literature, travel and forbidden knowledge and neglected science.

1. Space Travel:
William Burroughs was, in some ways, a science fiction writer, his books *The Soft Machine* (1961), *Nova Express* (1964), *The Ticket That Exploded* (1962) *The Wild Boys* (1971) and *Port of Saints* (1973) all contain elements drawn from the genre including Nova police and Nova mobs, time travel and an intergalactic war. *Nova Express* was even nominated for a Nebula Sci-Fi literature award.
But it is essential to understand that these are not simply speculative texts, but are instructional guides. Burroughs' literary style, especially his textual cut-ups and tape recorder experiments, offered him a way to escape from established literary forms, acting as devices that challenged the traditional structure of the novel. More than this, they also acted as devices employed by his protagonists within the novels, and, crucially, they offered potentialities for his readers as well. In *The Naked Lunch* Burroughs makes this explicit when he describes the work as a "blueprint, a How-To Book" (*The Naked Lunch,* p.222). The strategies suggested in Burroughs' novels became formalized in essays such as *The Electronic Revolution* and books such as *The Third Mind.*

Finally: "Man" Burroughs wrote, "Is an artifact designed for space travel. He is not designed to remain in his present biologic state any more than a tadpole is designed to remain a tadpole" (*The Adding Machine: Collected Essays,* p.82). For the author space travel was the only way to escape a world full of cops and agents of Control.

2. Beat Culture:

Burroughs was often considered one of the key beat authors, alongside Jack Kerouac and Allen Ginsberg. The three were collaborators and firm friends who had met in New York in the 1940s. Yet, despite the bonds of lifelong friendship, William Burroughs always distanced himself from the beats, his individualism marking him as a literary lone wolf, his geographical estrangement further separating him from his peers.

The beats offered an alternative and critical voice played against the conservative nature of post-war America. The culture of writers, artists and musicians that emerged in the coffee shops and jazz clubs of New York, San Francisco and beyond became a form of cultural rebellion that shed roots into subsequent counter-cultures. What William Burroughs shared with the beat writers was a belief in the fundamental right to personal freedom, an interest in the spiritual and in using aspects of his life for his writings. However, while Kerouac and Ginsberg wrote about their lived experiences directly, Burroughs was as interested in fantasies, the multiple states of consciousness and dreams as in the events considered day-to-day existence. He drew attention to the culturally constructed definition of consciousness in a letter to Allen Ginsberg where he asks the younger poet "Whose 'Normal Consciousness'?" before laying down Hasan i Sabbah's maxim "Nothing is True. Everything is permitted." (*The Yage Letters*, p.59). For William Burroughs there was no simple division between 'fantasy' and 'reality' this was a false either / or proposition.

Burroughs left the confines of the USA in 1949 and, other than occasional visits, he did not live there again permanently until January 1974, when he moved to New York City. The USA was a place he repeatedly condemned in his letters and works, and he preferred the freedoms offered by Mexico, Peru and Tangiers.

3. Pulp Literature:

Burroughs majored in English Literature at Harvard, graduating in 1936, and, while he had other dalliances with mainstream education, spending periods studying both anthropology and medicine, he never graduated from any other institutions. He was a fan of Jane Austen, but the book, which had a lasting impact on the writer, was *You Can't Win* by Jack Black, which he read in 1926 and which he would quote repeatedly throughout his life. The autobiography of a burglar / grifter / criminal, the book featured various figures who would subsequently appear in Burroughs' pantheon, including Salt Chunk Mary and the Sanctimonious Kid. The novel also introduced the young Burroughs to what would become one of his central concepts: the Johnson family (aka the Johnsons): "A Johnson" Burroughs would later write, "pays his debts and keeps his word. He minds his own business, but will give help when help is needed and asked for."(*You Can't Win*, p.v). As Burroughs wrote, the code of ethics offered by the Johnsons "made more sense to me than the arbitrary, hypocritical rules that were taken for granted as being 'right' by my peers." (*You Can't Win*, p.v).

Burroughs also had an interest in science fiction, when asked about his cut-up sources for *Nova Express* he states that it includes "quite a bit of science fiction" (*The Third Mind*, p.7). While in a 1974 interview with John Tytell, Burroughs stated that he read science fiction when he was able to "find any that's good."(*A Burroughs Compendium: Calling the Toads*, p.25). In addition he provided an introduction to the American edition of J G Ballard's *The Atrocity Exhibition*, reviewed Colin Wilson's *Mind Parasites* and wrote favorably about *The Mind Masters* by John Rossmann. Other pulp forms that he explored included the western novel, which in part informed *The Place of Dead Roads*. The form interested Burroughs when he returned to the Midwest, spending his final years in Lawrence, Kansas. The city is known as the site of a massacre led by Confederate guerrilla William Quantrill who attacked the town in 1863 killing an estimated 200 of the male residents. Burroughs, a believer in the right to bear arms, was careful to observe that the populace was not armed, and that the massacre was possible because "the mayor of the town for some reason impounded all the guns."(*A Burroughs Compendium: Calling the Toads,* p.44).

Burroughs' first book *Junkie*, (also, confusingly, issued as *Junky* and originally written as *Junk* but renamed at the behest of publishers) was published under the pen-name William Lee in 1953 as a 35c paperback by Ace Books, who doubled books, binding two titles in one volume, with a 'front' cover on each side. *Junkie* was paired with *Narcotic Agent,* Maurice Helbrandt's previously published novel. *Junkie* was the first of a proposed trilogy, the second volume of which was *Queer* (finally published in 1985, many years after it was first written); the third proposed volume was to be called *Yage.* The substance makes its first appearance in *Junkie* at the end of the book, with the narrator suggesting, "Yage may be the final fix." (*Junkie*, p.153).

4. Travel:

in an era before low cost flights and the *Lonely Planet* guides, travel meant something. Long distance travel necessitated people get on a boat or train, if they wanted to go somewhere off the familiar track often no one had mapped it. In a pre-Internet, pre-mass-telecommunication age international communication meant relying on postal networks.

Burroughs headed to Mexico at the end of the forties and to South America in 1951, before moving to Tangier in 1954. In 1951 and 1953 he searched the Amazon for Yage (aka Ayawaska), a plant that caused hallucinations, an experience of the deeper universe and was linked to telepathy. It is indicative of Burroughs' exemplary research and thirst for knowledge that he even knew the plant existed. Still largely unfamiliar it was certainly unknown in the 1950s. Burroughs' heard about it from the work of ethno botanist Richard Evans Schultes who was studying plants in Amazon in early 1940s. Burroughs was already on the case and his correspondence with Ginsberg regarding Yage - which detailed both of their experiences with the hallucinogenic vine - was published in 1963.

5. Forbidden knowledge:

alongside his search for Yage, William Burroughs also had an interest in Wilheim Reich and built an Orgone Accumulator in 1949. Even as Burroughs explored Reich's work, Reich was facing censure, seeing his work condemned and his books banned. Burroughs used an Orgone Accumulator throughout his life. Later the Dreamachine would be devised by Burroughs' collaborators Ian Sommerville and Brion Gysin, a device born from the exploration of flicker and its effect on consciousness, Burroughs would see in its spinning form new ways in which to challenge the restrictive forces of Control.

There is an antipathy towards authority in Burroughs' writing, and especially the mainstream science that introduced the atom bomb. Following the insight of his common law wife Joan Burroughs (nee Vollmer) he saw radiation from the atomic tests as having a detrimental effect on a psychic level. Here, even while avoiding biography, it is important to recall that as a youth Burroughs was sent to the Los Alamos Ranch School at Los Alamos. The school would later become the home of the atomic bomb.

Simultaneously to exploring unusual aspects of science Burroughs examined and embraced the magical, the forbidden and the fortean. He learned about magic from the rituals and dancers associated with the Master Musicians of Joujouka, from Brion Gysin, from experiments scrying while living at the Beat Hotel and from examining the effect of cut-ups, tape recorder experiments, scrapbooks and films. He became, later in his life, interested in the possibilities of Chaos Magick.

While not necessarily interested in established notions of numerology the recurring incidents of the number 23 fascinated Burroughs. It should be noted that Los Alamos was the scientific centre, but Trinity, where the first atom bomb was exploded was located 230 miles away. In a letter, dated May 23rd 1953, sent to Allen Ginsberg from Lima, Peru in which Burroughs' states: "enclosed a routine I dreamed up" - his first routine, a burlesque, scatological, satire called *Roosevelt after the Inauguration.* This form of writing would become the basis of the book he would start working on in Tangiers in 1956.

In the international zone of Tangiers, William Burroughs rented a room for $15 a month in the Villa Muniria (subsequently nicknamed the Villa Delirium) writing page after page, text piling upon text, spilling across the floor. "There were hundreds of pages of yellow foolscap all over the floor" Paul Bowles would recall, "covered with footprints, bits of old cheese sandwiches, rat droppings – it was filthy. When he finished a page he'd just throw it on the floor. He had no copies and when I asked him why didn't he pick it up, he said it was OK, it would get picked up one day"(*The Good Ship Venus: The Erotic Voyages of the Olympia Press*, p.251). In 'Burroughs in Tangier' Bowles described a damp room that opened onto a garden, one wall was pockmarked with holes from Burroughs' rigorous air pistol practice, and the other wall was plastered with photographs of Burroughs' trips in South America. In contrast to this description, the room's inhabitant remembered "a nice room… with a big comfortable bed." (*With William Burroughs: A Report From The Bunker*, p.24) Fuelled by joints and majoun, a sweet spicy cake made from marijuana, the author would write continually, producing much of the 'word hoard' that became *Naked Lunch* while living there.

Taking the title from Jack Kerouac *Naked Lunch* was at one time intended to be a combination of *Junk*, *Queer* and *Yage*. But it became something far more than that, as Burroughs would subsequently write it was the "frozen moment when everyone sees what is on the end of every fork." (*The Naked Lunch*, p.3). The routines that define the novel – camp, biting, cynical, acerbic, dryly humorous and dangerously percep- tive– came to Burroughs "like automatic writing produced by a hostile, independent entity who is saying in effect, 'I will write what I please.'" (*Letters 1945 - 59*, p.262) They spilled out as he hammered his way through typewriters in Tangiers and (later) Paris.

Years later Sonic Youth's Lee Renaldo would photo- graph a gutted corpse of a typewriter as it was absorbed by the rampant shrubbery and undergrowth of the aging author's garden in Lawrence, Kansas. Tossed aside and replaced, hammered into oblivion under Burroughs' fingertips, the corpse of the writer's machine. No computers. Learning to type was essen- tial for the writer, like learning to shoot.

1957, and Alan Ansen, Allen Ginsberg and Jack Kerouac all travelled through Tangiers, collating and typing the manuscript – helping it become more publisher friendly. Kerouac described it later: "I undertook to start typing it neatly double space for the publishers the following week I had horrible nightmares in my roof room - like of pulling out endless bolognas from my mouth, from my very entrails, feet of it, pulling and pulling out all the horror of what Bull saw, and wrote" (*Desolation Angels*, p.347). Later, Burroughs would inform his friend: "I'm shitting out my educated Midwestern background for once and for all" (*Desolation Angels*, p.347).

The Beat Hotel, a near flophouse where, so the mythology goes the bedcovers would be washed annually, was home at various times to Burroughs, Gregory Corso, Allen Ginsberg and Brion Gysin. It was Gysin who would over the ensuing decades become Burroughs' greatest influence. Located on the Left Bank, at 9 rue Git-le-coeur, the hotel was a short walk from the offices of Olympia Press, Maurice Girodias' legendary publishing house.

Under Girodias, Olympia published the Travel- ler's Companion series, numbered paperback books bound under the trademark olive green covers. The publisher was known for paying ex-pat writers and poets to produce salacious novels for 'travellers' who saw a freedom of the word in France that did not then exist in English speaking countries. Under fake names the likes of Alex Trocchi, Christopher Logue and others wrote sex and lust fuelled potboilers for the 'dirty book' market. Alongside these works Olympia published books that would subsequently become known as classics of modern literature, including

The Ginger Man by J P Donleavy, *Lolita* by Vladimir Nabokov, *Candy* by Terry Southern and *The Naked Lunch.*

There are a number of versions of how the manuscript came into the French publisher's hands, tales of the yellowed, rat chewed pages carried across Europe, of initial rejection and, following the publication of extracts in the American journal *Big Table*, eventual success. What matters is that William Burroughs, in his mid-forties, was paid $800 for the rights to the novel that had once been deemed as un-publishable.

The book was published in an edition of 5000 copies, number 76 in the Olympia Press Traveller's Companion series. Barney Rosset's Grove Press published it in 1962, and after legal battles it became readily available across America in 1966.

In November 1963, the UK, the work faced criticism and accusations of obscenity and pornography, and lengthy argument emerged in *The Times Literary Supplement* following the paper's publication of a review of *Dead Fingers Talk* a novel composed of sections from *Naked Lunch, The Soft Machine* and *The Ticket That Exploded.* These arguments flowed around notions of intention, interpretation and literary merit, perhaps the most succinct summary of the absurd nature of the criticism directed at the work came from Edith Sitwell who wrote "I do not wish to spend the rest of my life with my nose nailed to other people's lavatories. I prefer Chanel Number 5." (Appendix II, *The Naked Lunch*, p.264).

Those offering a defence argued that the book had social, cultural and literary merit, factors which become even more clearly apparent in the five decades since *Naked Lunch* was first published. However, elevating it to the status of *literature* - as if that was some special category - ignores some of the pleasures associated with that considered to be the low, the vulgar, the dirty, the obscene and the pornographic. To summarize the classic Burroughs' argument: *what literature?* When Sitwell divides literature between the 'toilet' and 'perfume' she creates an either / or

logic that is demonstrably false. As Burroughs would later state:
"Well, 'disgusting' doesn't refer to the books but to the subjective reaction of the person making the complaint. I don't think that anything is disgusting per se. These words 'disgusting' and 'filthy', etc., have prevented us from undertaking any scientific experimentation in sexual matters. How far would people get in physics if discovery was described as disgusting – 'Your formula is disgusting and filthy'? Not very far." (*Burroughs Live: The Collected Interviews of William S. Burroughs*, p.94)

In the base, in that toilet that so horrified Sitwell, in the filthy, in what the original reviewer in *The Times* summarized as "ectoplasm, jelly, errand boys, Ferris wheels, used contraceptives, centipedes, old photographs, jockstraps, turnstiles, newts and pubic hairs" (Appendix II, p.253) there is an undeniable truth that escapes ready closure. This list offered by the reviewer - with its apparently arbitrary and potentially endless litany of base, untouchable, and transgressive elements - recalls Georges Bataille's philosophical description of the heterogeneous, that which escapes the closure of homogenous definition. The heterogeneous cannot be simply defined, it pertains instead to the excess, defying ready classification and instead spinning into multiple possibilities. It belongs to the other, the space beyond the definable world; it emerges in moments of transgression. In engaging with the heterogeneous, Burroughs work exposed the forces of psychic control and ruptured the notion of order embraced by the homogenous world. In writing, in literature, the heterogeneous manifests itself most clearly in the pornographic, in the space in which language is forced to its limits. Burroughs forced language to its limits, in *Naked Lunch* he did this through the fragmentation of perspective, the collapsing of traditional narrative into a series of apparently arbitrary routines. Last, of course, he would attack the very structure and order of language, cutting into texts, folding texts, creating grids and (with Gysin) permutations. To recognize this excess is essential to appreciate *Naked Lunch*, a book which delves into the psychic quagmire, and the publication of which would help to free the words of authors.

Miscellaneous Notes & Lines of Flight.

There are many books on William Burroughs, the beats and related themes, and the following reading is merely indicative of what informed the current version of this text. The most essential works of course are those by William Burroughs and his contemporaries.

The book was first published as *The Naked Lunch*, although the title was actually *Naked Lunch*. By the mid 1980s most editions appear to have been published with the correct title.

There were numerous contemporaneous articles by beat writers and Burroughs published in men's magazines, see Jed Birmingham *Burroughs and Beats in Men's Magazines: William Burroughs Appearances in Adult Men's Magazines* at http://realitystudio.org/bibliographic-bunker/burroughs-and-beats-in-mens-magazines/william-burroughs-appearances-in-adult-mens-magazines/. For more on beatnik exploitation culture see Martin McIntosh, *Beatsville*, Outre Gallery Press, 2003.

For more on the vine Yage see William Burroughs and Allen Ginsberg's *The Yage Letters*, an absorbing study of Burroughs' search and the effect of the vine on both authors.

The notion of Burroughs as a science fiction writer and his influence on the radical edge of the genre is explored in Larry McCaffery (ed) *Storm the Reality Studio: A Casebook of Cyberpunk and Postmodern Science Fiction*, Duke University Press, 1991. The notion of the 'cop ridden planet' comes from an interview conducted by V Vale and Mindaugis with William Burroughs published in V. Vale and Andrea Juno, eds, *Re/Search 4/5 William Burroughs, Brion Gysin, Throbbing Gristle*, Re/Search Publications: San Francisco, 1982, p.21. See also, Allen Ginsberg, 'An Interview with William Burroughs' in William

Burroughs, *The Soft Machine, Nova Express, The Wild Boys: Three Novels By William S Burroughs*, Grove Press: New York, 1980, in which the author discusses the science fiction aspects of his work.

Not just a lone wolf of literature, as a child Burroughs authored a story entitled *The Autobiography of A Wolf*, truly the inaugural gesture of the first Wild Boy.

The notion of a wider aesthetically and culturally transformative beat culture is explored in Lisa Phillips et al, *Beat Culture and the New America: 1950 - 1965*, Whitney Museum of American Art, New York, Flammarion: Paris - New York, 1996.

For more on 'pulp' literature and its influence on Burroughs see Mike Evans, *The Beats: From Kerouac to Kesey, an Illustrated History of the Beat Generation*, Running Press: Philadelphia / London, 2007, p.14. See V Vale & Andrea Juno, *Re/Search: J G Ballard*, Re/Search Publications: San Francisco p.143 for the Burroughs preface to *The Atrocity Exhibition* and see Burroughs own *The Adding Machine: Collected Essays* for his writing on John Rossman.

Learning to shoot was essential for Burroughs, for more on weapons see William Burroughs *Painting and Guns*, which features an amusing interview on guns and tactics. Renaldo's image, focusing on the typewriter, serve to both emphasize the relationship of Burroughs to the machine - it is a tool, useful and readily replaced - yet simultaneously the machine is part of Burroughs, the two inseparable in the public imagination. In the David Cronenberg film version of *Naked Lunch* the typewriter becomes an insect, a muse and inspiration, a link to control and device to circumnavigate it. In Terry Wilson's book of conversations with Brion Gysin *Here To Go: Planet R101* Gysin describes Burroughs work with typewriters: "he was punching his way through a number of equally cheap plastic typewriters, using two very stiff forefingers… with enormous force. He could punch a machine into oblivion."(*Here to Go*, p.195)

Themes of travel run throughout beat literature and William Burroughs' letters offer a unique insight into the ex-pats survival strategies. Moving from celebrations of the local culture to letters asking for money and begging for simple communication they show the possibilities and risks that opened in travel. Collected and edited by Oliver Harris *Letters 1945 - 59* offers an essential guide to Burroughs' thought processes and philosophical and literary development.

See Oliver Harris's comments on Burroughs' letter to Ginsberg dated May 1st, 1950, for more on the psychic effects of radiation, in *Letters 1945 - 59*. Gysin and Terry Wilson discuss Burroughs, magic and spell casting in *Here: To Go: Planet R-101*. For more on Burroughs influence on contemporary magic see Genesis P-Orridge *Thee Psychick Bible*, Feral House: Port Townsend, 2009. Burroughs and Gysin's *The Third Mind* is essential reading for discussion and examples of cut-ups and cut-up experiments as is Burroughs *The Burroughs File*. Burroughs' aural experiments are discussed in Jack Sargeant 'William S Burroughs Primer' in *The Wire*, no. 300, Feb 2009, while the visual experiments are examined in Sergeant *Naked Lens: Beat Cinema*. For more on these experiments and the wider culture at 9 Git-le-coeur see Barry Miles' *Beat Hotel: Ginsberg, Burroughs and Corse in Paris, 1957 - 1963,* Grove Press, 2001 and the 'fictionalised' account in Brion Gysin's The *Last Museum*, Faber & Faber: London, 1986. Burroughs' interest in forbidden areas of knowledge inspired the fantasy of the Final Academy, the meeting of Wild Boys to be educated in these areas.

For more on the writing and publication of *The Naked Lunch*, as well as the history of Olympia Press, see John De St Jorre, *The Good Ship Venus: The Erotic Voyages of the Olympia Press*. Readers should also see Barry Miles' *William Burroughs El Hombre Invisible* Little Brown & Co, 1992 and Ted Morgan's *Literary Outlaw the Life and Times of William S Burroughs*, Avon Books, 1990.

Thanks to Lee Renaldo for permission to use his images of William Burroughs' typewriter, to Michael Spann and Marisa Giorgi for their ongoing inspiration and enthusiasm, to the MuMeson Archives, Sydney, where the first version of this lecture was presented, and to Roslyn and Miller, naturally.

Bibliography

Victor Bockris, *With William Burroughs: A Report From The Bunker*, Seaver Books: New York, 1981.

Paul Bowles, 'Burroughs in Tangier' in William S Burroughs *The Burroughs File*, City Lights Books: San Francisco, 1984.

William S Burroughs *Junkie*, Olympia Press: Paris, 1966 (1953)

William S Burroughs *The Naked Lunch*, John Calder: London, 1982 (1959).

William S Burroughs *Ghost of Chance*, Serpents Tail: London, 1991.

William S Burroughs, *The Adding Machine: Collected Essays*, John Calder: London, 1985.

William S Burroughs, *Ah Pook Is Here and Other Texts*, John Calder: London, 1979.

William S Burroughs 'Forward' in Jack Black, *You Can't Win*, Amok Books: New York, 1988.

William S Burroughs *The Burroughs File*, City Lights Books: San Francisco, 1984.

William S Burroughs *Letters 1945 - 59*, ed Oliver Harris, Penguin Books Ltd: London, 2009 (1993)

William S Burroughs *The Western Lands*, Picador: London, 1988 (1987)

William S Burroughs *Painting and Guns,*, Hanuman Books: New York, 1992.

William S Burroughs and Allen Ginsberg, *The Yage Letters*, City Lights: San Francisco, 1988 (1963).

William S Burroughs & Brion Gysin *The Third Mind*, John Calder: London, 1979 (1978)

Brion Gysin & Terry Wilson *Here To Go: Planet R101,* Quartet Books: London , Melbourne, New York, 1985

Neil Hennessy, 'Einstein's Brain Is Pickled In Kansas' in Denis Mahoney, Richard L Martin and Ron Whitehead *A Burroughs Compendium: Calling The Toads,* Ring Tarigh / Fringecore, 1998.

Jack Kerouac, *Desolation Angels*, Riverhead Books, 1995.

Sylvere Lotringer, ed, *Burroughs Live: The Collected Interviews of William S. Burroughs,* Semiotext(e): New York, 2001.

John De St Jorre, *The Good Ship Venus: The Erotic Voyages of the Olympia Press*, Pimlico: London, 1995.

John Tytell, 'An Interview with William S Burroughs', in Denis Mahoney, Richard L Martin and Ron Whitehead *A Burroughs Compendium: Calling The Toads,* Ring Tarigh / Fringecore, 1998.

Noah Weinstein and Randy Sarafan

THE PUSSYPAD

The Pussypad is a wearable haptic device that translates the realtime motions of female masturbation to video game control on the Atari gaming system. Like the Joydick, it is a fully functioning game controller; however, due to the different anatomical structures of the female genitalia, the Pussypad is composed of very different engineering. The Pussypad allows users to register 2-axis directional control via a thumb pad. Motion on the thumb pad is translated to a silicone clitoral stimulator, such that, cardinal directional control is derived from digital stimulation of the clitoris. Control of the fire button was mapped to the forward and back motion of sensors incorporated into a silicone penetrative device. With each stroke in and out of the vagina, the game player triggers one pressing of the fire button. For additional positive feedback, fire button presses result in short bursts of vibration delivered to the tip of the clitoral stimulator via an imbedded motor. The Pussypad is attached to the game player with a leather harness and rests comfortably between the legs.

The Pussypad 1.0 was revealed during a live performance of Joydick vs. Pussypad gameplay at Arse Elektronika in 2009 in San Francisco, and is currently undergoing version 2 design and engineering.

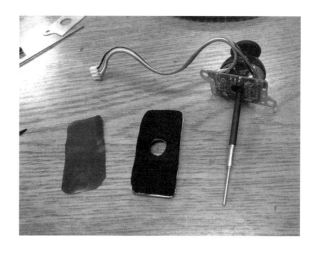

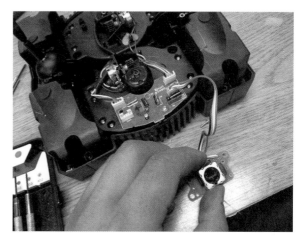

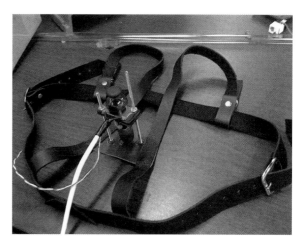

Christian Heller

GUROCHAN
IMAGEBOARD PERVERSION AND THE REINVENTION OF BODIES

Today, the internet is the medium that we invent ourselves in. It provides us with a growing choice of ideas, role models, and world views. Here we find identity resources to compose ourselves from, in a quantity and variety that was not available in earlier centuries. I like to talk myself into believing that this will lead to a collapse of mainstream normativity and a greater diversity in our cultures and identities.

If so, this would also have to hold true for our ideas of sex, gender and eroticism. Certainly: As vast and diverse a pornographic library as can be roamed online never existed prior in human history; nor was it ever so easy to find material, information and communities for even the most remote fetish. The internet joke "Rule 34" refers to this phenomenon stating: No matter what circumstance or object you can think of, there will be porn of it somewhere on the internet. Ergo: Any conceivable sexual identity, desire or idea will find affirmation and exploration somewhere on the web.

Sexual identity resources abound on the internet, in many forms and many backgrounds: porn paysites for any fetish that might attract solvent customers; sex education sites for adolescents and adults; adult-only 3D play areas on Second Life; dating sites like OkCupid.com and more narrowly specialized communities like FetLife.com or the German GayRomeo.com. They all provide infrastructure, inspiration and reinforcement for our sexual imagination and practice and our conceptualization of the human body in its abilities and its ideal form. As we face a growing technical ability to alter not only our virtual bodies on places like Second Life but our physical bodies as well via operations like sex changes or breast enlargements, growing body image creativity on the internet can be expected to extend into the outside world and maybe even society at large.

In this text I want to focus mainly (with one major, necessary digression) on one specific corner of the internet and how it provides resources for an expanding imagination of sexual desires, gender and, in particular, bodies: a site called GUROchan.

I consider it to lie rather far away from the center of mainstream sexuality and porn, focussing on some extreme fetishes with not much backing or sympathy by that mainstream. To some, the contents of GUROchan might qualify as one of the dark corners of the internet best kept in obscurity, probably one of the things that ought to be censored if internet censorship was possible. In contrast, I want to put forward an argument on how we can consider it as a progressive outpost on the widening frontier of how we define ourselves on the internet in questions of sexuality and body.

Imageboards as a pornographic environment

GUROchan is an internet imageboard site: a sort of web forum for posting and discussing images. It shares its form and manners with several other websites that also have a "chan" in their name or URL, like the popular Japanese sites 2ch.net ("2channel") and 2chan.net, their English-language counterparts 4chan.org and 7chan.org and the German-language krautchan.net.

On all of these sites, content is organized into a number of different sub-sections: imageboards devoted to special topics or content forms. Browsing through such an imageboard, one will see a chronological list of threads. Such a thread consists of a linear chronology of postings of images and/or text. If the thread is short enough, it will be shown complete on the imageboard index of threads; if not, the oldest, first posting and a few of the most recent postings will be shown. One can open the thread in a separate page to see it in its full length.

Imageboards usually provide very low barriers for posting content, for opening a thread or for appending something to one: No user registration is needed. Any browser view of an individual imageboard or thread already contains the input fields for creating a posting. Anonymous participation is not only possible but mostly the norm. Postings are published instantly, without previously going through any content control. They can be deleted subsequently. The threat of subsequent deletion and the banning of individual computers or regions from posting material is the only instrument of content regulation available to imageboard moderators.

This provides an interesting playing field for porn!

Imageboard content is content posted by the imageboard users: There is no gap between those who post and those who consume. Posted material includes self-made material as well as material taken from anywhere else, photographed, scanned or simply copied from some other website. Content is not filtered by an editorial selection streamlined for purposes of teasing someone into buying a paid membership as on a porn paysite; it does not go through a bottleneck of market optimization. Nor is it limited by the means of production, the artistic expertise or the actor bodies available to those individuals who want to provide imageboard input: If they are unable to compose themselves the quality content that they would like to provide, they can just provide amateur or professional material found somewhere else on the internet. Naturally, imageboards would collapse under the scrutiny of copyright laws if those were strictly and effectively enforced.

As a result, porn found on imageboards is selected and filtered very directly by the desires and ideas of their audiences, not by what paysite owners think they can make money with. Nor are these desires and ideas of those who post, request and comment on images limited by the self-control of shame or saving face: The participants are anonymous, so why should they care? Additionally, on many imageboards there is an expiration date for threads: They get deleted after a certain while. This ephemeral nature eases pressures of copyright and other laws: Any picture copied illegally from somewhere else will have disappeared before the lawyers of the copyright owners had the time to identify them and write a nasty letter to the site operator.

Imageboards provide few obstacles to a shameless and open performance of their users' interests. As a result, sex and porn are one of their major subjects. To better estimate GUROchan's position in this environment, lets first take a look at the pornographic mainstream of a far more popular English-language imageboard site: 4chan. English-language imageboard culture often defines itself (positively or negatively) in relation to 4chan, and so does GUROchan. As a bonus, we will learn about some terms and phenomena that will become very important when discussing GUROchan.

A short look at 4chan's porn culture

4chan.org is a very large imageboard site, carrying, as of writing this in June 2010, forty-nine mostly highly-active imageboards devoted to a variety of subjects ranging from "Video Games" and "Papercraft & Origami" to a political "News" board. A lot of the boards carry Japanese names and refer to Japanese popular culture. This is where imageboards originated from, and affinities towards Japan are still extremely strong on places like 4chan, as we shall see in the content and technical terms used.

Thirteen of 4chan's boards are assigned an "18+" age warning. Six of these adult-only boards carry names that designate them exclusively for various genres of sexually tantalizing content (as is usual on such sites, individual boards carry a cryptically short label of one or a few more letters, mirroring their URL sub-folder, abbreviating their purpose): "/s/" ("Sexy Beautiful Women") claims photographic content. The other five contain drawn images in manga style, differentiated into several genres: "/y/" ("Yaoi") depicts male homosexuality and "/u/" ("Yuri") depicts female homosexuality. "/e/" ("Ecchi") is for less explicit and "/h/" ("Hentai") for more explicit material. There is also "/d/" ("Hentai/Alternative"), about which I will be talking in just a minute.

The other half of the "18+" boards is differentiated according to more technical criteria: for posting images of a certain filetype or size ("/hr/": "High Resolution", "/gif/": "Animated GIF"), for linking outside downloading and file sharing locations ("/rs/": "Rapidshares", "/t/": "Torrents") or specifically for describing and requesting certain images in the hope that some reader has something fitting on their hard drive and will post it. These boards do not trade exclusively in sexual material, but it does make up a large chunk of their content. The most popular "18+" board though is also overall the most popular board on 4chan: "/b/" ("Random"). It's the anarchy board: Any content is allowed that is not spam or illegal according to United States law. Naturally, sexually explicit material pops up here as well. In relative numbers, it's not as large a chunk of board content as on the designated porn boards; but in absolute numbers, it can hold up to them easily, due to "/b/" alone accounting for a third of all the traffic of all of 4chan.[1]

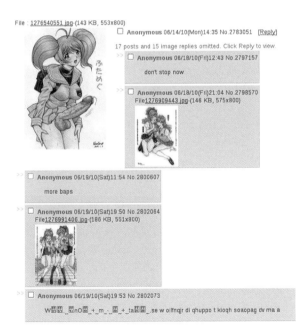

Example of an imageboard thread: A thread of pictures depicting "futanari" on the /d/ board of 4chan.org

What kind of sexually explicit and pornographic material gets posted on /b/? Due to several new threads starting any minute and extremely short thread expiration times (it's common for threads to

disappear in less than one hour), it's a little difficult to find a stable quantity to count. For a quick overview, I decided to reload the first index page of the "/b/" board ten times in a row, with a few minutes pause in-between, and to count what material posted I'd see under these circumstances. I got:

- 2 screenshot series from commercial hardcore male-on-female sex videos with the URLs of paysites to download them from superimposed
- 2 picture series of teenage girls undressing in front of their webcams
- 1 webcam picture of a person looking very female (apart from having a penis) masturbating another person looking male
- 1 high quality photo, looking professionally staged, of a male penetrating a female, with a paysite URL superimposed
- 1 photo of a nude child in a sexualizing pose, with her vagina hidden under a picture of "Pedobear", the manga-style pedophile mascot of 4chan
- 1 photo of a penis bended backwards and thereby inserted into the anus of its owner, with the words "GO FUCK YOURSELF" superimposed
- 1 professionally looking photo series of nude males
- 1 unprofessionally looking photo series of a woman inserting a banana and a beverage can into her vagina
- 1 series of photos of children scantily dressed, looking like professionally staged erotica and with URLs superimposed
- 1 amateur-like looking photo of a white female engaging in a sexual act with two black males
- 1 (manipulated?) photo of two women sucking on something that starts as one penis but forks into two penises in the middle of its length
- 3 series of photos of penises getting masturbated by their owners
- 2 professionally looking erotic nude pictures of females
- 1 series of manga-style drawn images of small children in sexual poses, nude or scantily dressed

- 1 series of amateur photos of nude females and males
- 1 photo of an old woman showcasing her breasts
- 1 photo of a female, scantily dressed in the colors of the American flag, holding sparklers in her hands and showcasing some kind of firework protruding from her vagina
- 1 professionally looking photo series of a young teen posing naked

Admittedly, this is a very small sample for judging the pornographic landscape of /b/. I will try to keep this limitation in my mind and express the following only as a first dirty quick shot: The users of /b/ appear interested not only in pornography focussing on the female body, but also in seeing penises and male-only erotica. Material from paysites seems to get posted just as much as amateur material, often recorded directly via webcam. There appears to be some interest in sexually abnormal anatomy and action. There also seems to be some interest in material concerning non-adults, most of which borders illegality by most western laws.

4chan has a page dictating rules for its site and its individual boards[2]. It explains the orientation of the individual boards and according to what criteria material gets deleted or users banned. It declares 4chan to be subject to United States law and its limitations in what content can be allowed: Therefore, images most of the western world would consider "child porn", at least photographic ones, are forbidden anywhere on 4chan, including /b/. Moderators delete them if they see them, but it's hard to keep a firm and always instantaneous grip on the raging pandemonium that is /b/, so they do appear and occasionally live on for short times before being exterminated.

The pornographic anarchy just described mostly ends at the border of /b/. Point 3 of the "Global" rules reads: "Do not post the following outside of /b/: [...] anthropomorphic ('furry'), grotesque ('guro'), or loli/shota pornography." Let's decipher this!

The ban of "loli/shota" refers to "lolicon" (the etymology to *Lolita* is obvious) and "shotacon", i.e.: to erotic images of female and male children drawn in manga style. Since these are only banned in the rule referring to parts of 4chan that are not /b/, we can assume that this decidedly pedophile art form is somehow tolerated in /b/. Considering the general damnation of and intolerance towards any kind of pedophile erotica that currently permeates western society, this suggests an interesting degree of permissiveness in the morals of 4chan's site operators.

The ban of "furry" pornography might need some more explanation. "Furries" are fictional creatures that exhibit characteristics of humans as well as of non-human, furry mammals. They are often depicted in a cartoony way, reminiscent of anthropomorphized animals with human intelligence known from fairy tales or Walt Disney's animated shorts. There is an entire subculture devoted to them: people meeting at furry conventions and wearing furry costumes, sometimes developing elaborate new "furry" identities for themselves. To some, the furry lifestyle also involves furry eroticism. So yes, there is "furry pornography". But this all sounds rather harmless. What's so bad about "furries" that they should be banned from the imageboards of 4chan the same way that pedophile pornography is banned? We will return to this question later in the text when we deal with GUROchan: Furry content is banned there as well. Actually, imageboards without a negative attitude towards furry content seem to be the exception rather than the rule, even if they allow for other material most of the mainstream public would consider highly deviant.

The third ban deals directly with the term that GUROchan is focussed on: "guro" pornography is also banned outside of /b/. The rule tries to explain the banned category as pornography that is "grotesque". That's not really enlightening. We will deal with a definition of "guro" pornography later when we take a closer look at GUROchan.

Let's examine the adult-only boards designated for sexual content. We will focus on the five boards for drawn material, as this is the area that GUROchan also deals in, mostly ignoring /s/ ("Sexy Beautiful Women"). (It contains photographic material that ranges from portrait and nude photos to softcore porn, often taken from paysites. Nothing hardcore or "grotesque" though.)

The most innocent of these boards, hardly worthy of being called pornographic, is /e/ ("Ecchi"). The 4chan rules for it state that "[o]nly softcore nudity is permitted", and it's hard to find anything there that goes beyond drawings of normal human females looking cute and posing a little, maybe naked. A quick glance at the homosexual boards /y/ ("Yaoi") and /u/ ("Yuri") reveals standardized cute cartoons characters (sometimes borrowed from pre-existing fictional universes) engaging in standardized romantic and sexual situations. Occasionally, some minor non- or super-human trait may appear, such as otherwise human-looking characters having cat ears or an unusually large penis. These boards appear to be interested in one narrow subject and few beyond. The /h/ ("Hentai") board claims to be more general-purpose: the rules allow for manga-style drawn porn images that do not fit on the other porn boards. Here, a lot of heterosexual hardcore imagery of sex acts is to be found, often with exaggerated anatomical possibilities; but they seem to pose no danger to the "no grotesque!" rule, their occasional unrealism venturing not beyond what is to be expected of bending human behaviour, anatomy and potency a little to emphasize features usually lusted for in mainstream pornography.

Then, of course, we have /d/ ("Hentai/Alternative"). The rules for this board sound interesting indeed: "'Alternative' images including: futanari, bondage, tentacles, etc. are welcome." "Bondage" needs no explanation. "Tentacle" refers to "tentacle porn" which explores the stimuli of tentacles (belonging to real or fictional classes of creatures) on the human body, a subject in Japanese culture for at least two centuries[3]. "Futanari" is a genre depicting otherwise female bodies with male genitalia. That's just it: It does not necessarily entail a strict concept on how these genitalia get there, whether by birth as a hermaphrodite, by operation or through magic; or what this means for the sexual identity of their owner, or for the sexual identity of the viewer jerking off to male and female attributes at the same time.

Browsing through /d/, "futanari" is the main fetish to be found, but the explicitly allowed tentacle and bondage pornography appears also. In general, compared to the other boards (apart from /b/), a greater permissiveness for the bizarre appears to exist. Sexualized non-human creatures such as mermaids, fairies or reptile girls appear regularly (but, alas, no furries!). Further bodily abnormalities occur, such as extreme proportions or multiplied genitalia and extremities.

But the rules for /d/ ramble on: "Images depicting bestiality, guro, scat, or generally seen to be 'extreme' in nature are not to be uploaded." So, sex with animals is not allowed (what about mermaids and reptile girls?), nor is playing with feces and urine. Banning "guro", already banned in one of the "Global" rules for any board that is not /b/, is a redundancy. Maybe this extra emphasis is necessitated by the greater affinity towards the bizarre, which more strongly threatens a seduction towards the "grotesque". The danger seems to be dependent on the dose: Nothing "'extreme' in nature" is allowed.

What could a more extreme dosage look like?

Enter GUROchan

Apparently, there once was a /g/ board for "guro" on 4chan. It got deleted (and the label "/g/" repurposed for another board)[4]. Around the same time, GUROchan came into being, without a doubt catching some of the fleeing activity now banned in 4chan. So let's take a look at this GUROchan!

What "grotesque" content does GUROchan specialize in? Let's ask the site's FAQ:

"'Guro' is a shortening of 'ero guro' which itself is a shortening of the term 'ero guro nansensu'/'erotic grotesque nonsense' and originally referred to an artistic movement that took flight in the early 20th century, Japan. The modern definition of guro has evolved slightly since then as it has worked its way into different media but still stays true to the original concept.

Guro does not refer solely to gore, rather it is an umbrella term that encompasses a wide variety of frequently erotic, artistic deviancy often with negative subject matter such as, but not exclusive to, death. Guro is an exploration of the depths of extreme fetishism and deviant hedonism."

Okay, that sounds a bit like something 4chan's site operators could have had in mind when defining the extremes not wished for outside of /b/.

Let's follow the reference to "ero guro nansensu" a little closer. Wikipedia helpfully has an English-language article on the term "Ero guro".[5] It talks about "a literary and artistic movement from 1920s and '30s Japan" with a "focus on eroticism, sexual corruption and decadence". Further keywords thrown around are a parallelization to the atmosphere of Weimar Republic Berlin, early influences in Japanese woodcuts depicting violence and mutilation from the 19th century and the contemporary "Sada Abe Incident" involving a consensual sexual murder and castration (known to the western public mostly through Nagisa Oshima's movie "In the Realm of the Senses"). Further, this complex is said to have influenced post-war Japanese culture, porn and horror movies, manga and anime and even pop music.

It's interesting that GUROchan's FAQ entry on the term "guro" needs to stress that it "does not refer solely to gore". This term for graphic violence with an emphasis on protruding blood and guts seems to have caught on as one popular translation of "guro" in English-language imageboard culture, no doubt due to its close phonetic similarity, whereas "grotesque" sounds much broader in its possible meanings. On GUROchan, we will find both readings translated into pornographic matter.

Quick content overview

Just like 4chan, GUROchan offers a number of imageboards specialized in different types of content, defined by a set of rules. In contrast to the example of 4chan, it also includes two text-only forums and archives for some of its sections. GUROchan apparently is not interested in regularly erasing all of its material.

The text forums are "/dis/" ("Discussion") and "/lit/" ("Literature"). /dis/ is a general-purpose discussion forum for GUROchan's users whereas /lit/ is specifically for erotic story writing.

There are three topical imageboards: "/s/" ("Scat"), "/g/" ("Gore / Death") and "/f/". I will talk about each of them in greater detail below.

Then there is the "/req/" board for "Requests" (which works like 4chan's "Requests" board: ask for some individual image or a class of images and hope that somebody posts them) and "/p2p/" used for trading file sharing links. "/art/" ("Artwork") and "/kaki/" ("Oekaki") are specifically for original art drawn by the GUROchan users themselves, with the latter offering a Java paint applet.

GUROchan's imageboards are for drawn content only. GUROchan's rules [6] specify it:

"NON-DRAWN IMAGES (3D renders and photographs, photo-shopped or otherwise) ARE NOT ALLOWED ON ANY OF THE BOARDS."

Guro scat

The "Scat" board /s/ is the most narrowly focused in subjects depicted. Allowing for "shit, piss, vomit, and related artwork" according to GUROchan's rules, it mostly deals in the first.

I browsed the last fifty threads available at the time of writing this and counted what was being depicted in their respective opening drawing:

- 25 pictures dealt exclusively with ordinary human beings defecating, all except one one of them female; in one case the female was simultaneously masturbating; in another case, a diaper was involved.
- 6 pictures dealt with the experience of being defecated on: in one case by a female giant on a miniature male; in another, two females defecating into each others mouths in a sixty-nine position.
- 4 pictures dealt exclusively with females sitting on the toilet, not showing the outcome.
- 4 dealt exclusively with humans urinating, one of them male.
- 2 further pictures depicted simultaneous defecation and urination, once by a female and once by a futanari.
- 2 pictures depicted females being fed their own feces through a tube going from anus to mouth.
- 2 pictures showed females bathing in feces.
- There was one further case each for a male vomiting, a female farting, a female getting an enema, a female simultaneously urinating and drinking urine and a female carrying her feces in a bale of cloth.

Notice that in all but five cases, only human females were depicted. Most images show only a single person.

If we accept a fetish for feces and urine, most of this material is free from surprises and hardly "grotesque". But of course, a fetish for feces and urine is hardly accepted as normal in the mainstream. To quote a discussion from the discussion board /dis/ entitled "What started the trend of grouping Scat with Guro?" [7]:

"Guro is short for the Japanese version of 'grotesque'. Anything someone might consider gross falls under the title, I guess."

Compared to the other boards, /s/ is not very inventive. In the overview concerning its contents given above, a few bizarre elements pop up peripherally that we will find much more fleshed out on the other boards. Compared to the varieties of the /g/ ("Gore / Death") and /f/ ("Freakshow") boards that we are about to explore, the scat board /s/ looks like one very specific fetish discretely packaged into isolation.

Guro gore

The "Gore / Death" board /g/ allows "for gory images", according to GUROchan's rules. All images to be found here deal with at least one of the following: bodily violence, killing, corpses, mutilation and disintegration of bodies. The context is often sexualizing. In contrast to the "Scat" board /s/, the variety of situations depicted is great. A quick browsing through the most recent imageboard index pages reveals individual and sometimes multiple threads about ...

- animals mauling humans to death
- different methods of asphyxiation: hanging, bagging, drowning etc.
- raping babies, tearing them up from the inside through penile penetration
- the digestion of females inside of animals and monsters, as seen from the inside of the stomach
- disintegrating bodies by melting them
- torturing breasts with needles and nails and by cutting into them
- beating people with fists
- castrating males with sharp objects
- "neck fucking", that is: penile penetration of the neck after the head has been cut off

- males being prepared as food, served on a plate, cooked or roasted, cut up and eaten
- "Zombie girls", sometimes eating parts of humans
- breaking spines
- empty eye sockets
- hearts getting ripped out of the ribcage
- giving birth in a gory and often deadly way
- miniaturized persons getting crushed, eaten or impaled by penises
- pushing penises into non-oral, artificially created openings in the head, entering the brain
- execution by gun or rifle
- different methods of decapitation
- females offering their breasts sandwiched into hamburgers
- collecting different ideas about torturing and mutilating specific persons (like popular fictional manga and anime characters)
- intestines and anatomy with an almost medicinal interest

Typical GUROchan /g/ artwork, posted by "Anonymous".
Thread URL: http://gurochan.net/g/res/39.html

Drawn subjects range from minor injuries and pains to total murder and destruction; from sanitized actions keeping the body intact at least superficially to a total splattering of the body into a large number of small parts, a pool of blood and guts; from realistic scenarios that could happen in everyday life to fantasy and science-fiction environments with elaborate machines, monsters and magic.

Sometimes, the sexualized nature of the violence is emphasized, with beautiful and attractive (in accordance to mainstream tastes) victims, often naked, being subjected to genital interaction. But the violence and bodily disintegration can also take the foreground to a degree that may be arousing to a certain fetish of violence, but hardly stresses the sexual experience of the victim or protagonist, does not involve overtly sexual activity and does not allow for the admiration of primary or secondary sexual characteristics (that are drowned in blood and intestines), up to making the identification of the sex of the victim impossible. Explicitly non-sexual scenarios are as frequent as sexual ones, such as a fight to the death between enemies or politically motivated executions.

It's primarily the female form that is getting mutilated, just as often by males as by females; very often, the perpetrator is not visible at all. The tables are turned occasionally: Males appear as objects of gore and death fetishes more often than they did in the previous sample of /s/ as objects of the scat fetish. In general, penis torture, mutilation and castration seem to be a popular subject, but it strikes ordinary males just as much as the hermaphroditic "futanari" population.

Guro freak

The /f/ board ("Freakshow") has the least strictly defined topical content according to GUROchan's rules: "A general board for very weird stuff that isn't too bloody, or perhaps not bloody at all." The FAQ of GUROchan is a little more elaborate:

"/f/reakshow is intended to be a board for extreme, weird and otherwise lurid material that does not involve scat or grievous wounds. To put that in context, imagine /f/ as an extreme version of 4chan's

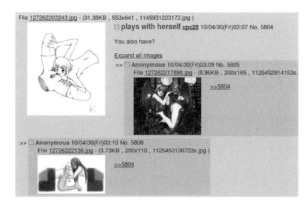

Typical GUROchan /f/ thread.
Thread URL: http://gurochan.net/f/res/5804.html

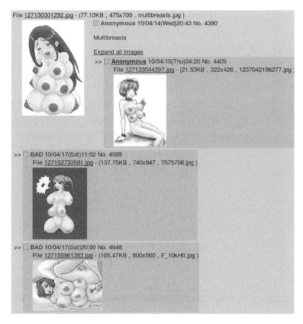

Typical GUROchan /f/ thread.
Thread URL: http://gurochan.net/f/res/4390.html

/d/ (hentai/alternative), and while the two may overlap on some subjects (for example, monster girls) redundancy is not intended. There is absolutely no need to have two boards that cater for exactly the same thing; this is why something such as bondage does not belong here."

In other words: /f/ is for anything "weird" that is too extreme for 4chan's /d/ (remember that /d/ also initially allows for a great variety of "alternative" stuff,

but quickly limits this by banning anything "extreme") and is not already claimed by the other two topical GUROchan boards. (Bondage, apparently, is not unusual enough, at least not in its common form.) And we will actually find the most imaginative parts of GUROchan's pornographical landscape here.

So let's take a stroll through /f/ and collect some impressions. Here's a number of thread subjects:

- "Unbirth", that is: people crawling full-body inside a living female, by way of her vagina
- images of anorexic persons and semi-skeletons
- incomplete bodies (lacking, for example, a head or a torso) that are alive and don't seem to find their state unusual or painful
- "Parasite dicks": creatures entering the vagina of females with a part of their body protruding that looks like a penis
- flexible elongated objects (tentacles, chains) non-mortally piercing bodies from anus or vagina up to the mouth
- bestiality
- inflation of bodies by, for example, pumping air or liquid into them
- amputees, not in the (gory) act of getting amputated, but actually living their daily life lacking limbs
- conjoined twins
- x-ray transparency of bodies getting penetrated, showing the interior organs' reaction
- male pregnancy
- living persons becoming inanimate
- girls being kept in tubes
- persons with more arms or legs than is usual
- multi-breasted women
- mermaids and their male counterparts
- *iants and miniature persons
- non-humans: human-animal hybrids, robots, monsters – and having sex with them
- futanari, of course, and their counterparts: otherwise male-looking humans with a vagina instead of a penis
- albinos

- people turned into objects: not just pretending, but actually being transformed into them
- girls not just growing penises, but actually being transformed into them whole-body

As is to be expected, /f/ is much less gory und much less filled with deadliness than the "Gore / Death" board /g/. Visible non-consensuality in the scenarios depicted also is less frequent. Nevertheless, the amount of bodily transformation from a normal state to an abnormal state is just as large and extreme as on /g/, by realistic means, but also to a much larger degree by science-fictional or magical means.

The realistic part of the material deals with existing medical conditions (like genetic abnormalities) and body modification procedures (like amputation) that create bodies different from what is considered normal or healthy. The unrealistic part either bends the aforementioned causes into physically and biologically questionable extremes (sure, we can try to push larger and larger objects into a vagina, but a person being entered in full by another grown-up person this way could hardly survive the way she does on /f/). Or it just invents an implausible new form by extrapolating not from nature but from art and literature, symbolic imagination, magical realism and anthropomorphization of fetishes that originally don't deal with full human beings.

Human females in a form to be expected as normal from conventional heterosexual pornography are a minority on /f/. It makes more sense to speak of a prevalence of human female as opposed to human male characteristics in a crowd of creatures of often debatable affiliation in species and gender. The boundaries get fuzzy here.

Guro demographics

It's easy to get lost in a de-personalized view of an imageboard where most postings give "Anonymous" as their author. Sites like 4chan have evolved this phenomenon into the idea of a collective uber-personality composed of the constantly changing and volatile crowd of its users, re-personalized by carrying a common mask face and by speaking in a common (computer-synthesized) individual voice.

But on GUROchan it's actually not so unusual to find individualized pseudonyms as authors instead of the common "Anonymous". For example: On the boards for self-made artwork, /kaki/ and /art/, individual artists will sign postings of their work with their nickname. Still, "Anonymous" seems to be the majority in sheer number of postings on GUROchan.

But this does not mean that we cannot explore a bit the demographics of this site. We only have to trust the self-descriptions given in numerous survey threads on the discussion board /dis/. I will here evaluate one of the most comprehensive ones, entitled "Guro Survey!", from November 2009, making things easy for me by a consistent form asking for attributes like age, physical sex, sexual orientation and fetishes. [8] 71 answers following directly the order of the survey form were given. The conditions of such an imageboard survey and the small sample size are admittedly scientifically weak, but they will have to suffice for a first approximation. Let's look at the individuals thus characterized.

The first question was for age. Four of the seventy-one participants refused to give an age. For the remainder, the average age is 20 years. Here are the details:

- Let's start with the youngest. 1 participant answered fifteen years of age, 2 answered sixteen, 4 answered seventeen, 14 answered eighteen, 11 answered nineteen and 8 answered twenty years. Thus, a clear majority of forty out of sixty-seven answered 20 years or younger.
- The next age group ranges from 21 to 27 years of age: 2 answered twenty-one, 3 answered twenty-two, 7 answered twenty-three, 3 answered twenty-four, 4 answered twenty-five, 2 answered twenty-six and an additional 2 answered

twenty-seven years. This age group contains twenty-three out of sixty-seven, almost a third.

- One each answered thirty years, thirty-four years, thirty-five years and thirty-eight years. The 30+ years age group counts only four members.

So, GUROchan seems to attract used mostly young people, at the end of their teens or in their early twenties.

What is the male-female ratio and the distribution of sexual orientations? Almost all participants gave a clear indication of their biological sex; one claimed to be intersexed and one claimed to be androgynous, but both still put the attribute "female" as their final word on the matter into the input field and will for the purposes of this evaluation be counted as such. Thus, we get 38 males and 33 females. In regard to their sexual orientation, 30 of the males claimed to be sexually interested in females only. 2 claimed to be homosexual and the remainder of 6 answered either "pansexual" or "bisexual". The distribution is a little different in the female camp: Here, only 9 claimed to be interested in men alone. 6 claimed to be interested exclusively in females, whereas 16 claimed to be either "bisexual" or "pansexual". Additionally, one female claimed to be "queer" and one claimed to be "asexual".

Going by the survey, it looks like GUROchan has a relatively balanced male-female audience. The male part though appears much more exclusively heterosexually oriented (four out of five) than the female one (one out of three). This resonates well with the ratio of many females vs. few males depicted as pornographical objects on the imageboards.

It's much harder to quantify the predilections towards different fetishes given in the survey, as the categories seem to be drawn very individually and most answers consist of long lists of very specific images and fantasies that are hard to group together over the whole number of answers. A few interesting trends do appear, though: As is to be expected, fetishes and situations that would be grouped into the /g/ ("Gore

/ Death") and the /f/ ("Freakshow") boards are very popular. Scat material is given relatively rarely as a "Turn On"; it mostly appears as a "Turn Off". Looking at single fetishes in isolation, we can also find out interesting things like a fifth of the survey takers claiming to be turned on explicitly by pedophile "lolicon" art, which is much more than the share of people liking "futanari" imagery. But for more meaningful evaluations of GUROchan's sexual landscape, we will have to look beyond mere enumerations of single fetishes.

Fetish wars or tolerance?

The aforementioned survey differentiated between "Turn On" and "Turn Off" material, and it was not unusual for some fetish to appear as one person's turn on and at the same time as another person's turn off. As we have seen on 4chan, specialized fetish boards often do not like to have some material intrude from another fetish universe, with the degree of permissiveness different from board to board. How does this dynamic play out on GUROchan, a place that seems specialized in outcast fetishes banned on most other imageboards?

As its rule set shows, GUROchan is not a place where absolutely anything is allowed: It has to resonate with certain aesthetic sensibilities. I already quoted the global ban on non-drawn images as an example.

There is also a more or less strict content segregation between individual boards: Bloody stuff belongs into the "Gore / Death" board /g/, non-bloody "weird" stuff into the "Freakshow" board /f/. In practise, these two often overlap in situations, bodily deformations and fetishes depicted. Their common relationship to /s/ is more interesting: The "Scat" board has few common themes with /f/ and /g/, and as we have seen in the survey, scat appears to be a turn off for a large part of the GUROchan crowd. They are happy about it being segregated away into its own board. The general opinion, affirmed by various discussions on /dis/, seems to be: We accept your presence as there is hardly any other place you people can go to; but please stay on your own board!

Compared to this segregation, the lines between turn-on and turn-off material as shown in the survey is much too erratic to be mirrored by the dividing line between /f/ and /g/. On either board, but on /f/ especially, a visitor has to endure extreme examples of a wide range of erotic universes that are probably very far away from what this visitor came for. It's the fate of GUROchan's outcast fetishists to learn tolerance. If you want to see women beaten violently on /g/, you will also occasionally have to endure pictures of males getting their testicles crushed. If you browse /f/ for building a collection of extreme futanari porn, better prepare for also getting some bestiality thrown in your face.

I believe this to be a good thing: It's the opposite to pornographic normativity promoting narrow ideals for bodies and sexualities. It widens the individual's horizon on what is possible erotically and on the diversity of people's attractions. Confronted with so much different input, one might even find oneself drawn to subjects outside of one's sexual self-image, provoking its redefinition.

Given this dynamic, though, the common GUROchan disgust towards "scat" may provoke mystery. A more extreme example, though, is stated in the following global GUROchan rule:

"Furry pictures and overly simplistic 'toons' are not welcome here."

Racism towards furries?

I have already talked a bit about the furry world in my observations regarding 4chan. 4chan bans furry material just as it bans "guro" and "lolicon" (which, naturally, can also be found on GUROchan, as long as it is somehow guro-fied). Given that GUROchan bans furry material as well, this must surely be the worst outcast fetish ever! But it's hard to imagine a specific disgust towards porn of anthropomorphized

furry animals in a crowd that tolerates much more grotesque material, including non-furry anthropomorphized animals. Furry material turned up as a "Turn off" on the GUROchan survey above a few times, just like "scat" did; but not more frequently than any other fetish generally tolerated on GUROchan and appearing just as often as a "turn on".

Diving a little into the discussions concerning the ban on furry material on GUROchan's /dis/, a view of this ban as somewhat ironical given GUROchan's general fetishistic openness can certainly be found. The discourse then usually takes a turn towards an argument that sounds a little more plausible as a cause for a ban. Here is an example:

"Furry as a sub-genre of anthro is basically /f/. It's just that there's a huge fuckin' fagdom behind it and every guro community tries to stay as far from it as possible to avoid drama and toony furspam." [10]

So it's not so much the erotic nature of furry porn that GUROchan's community wants to avoid, but the social implications of letting it in. Apparently,

Example of "Furry" art, taken from Wikipedia's page for "Furry fandom". Author: "Yamavu" See http://en.wikipedia.org/wiki/File:Anthro_vixen.jpg for more details.

non-furry imageboard contact with the furry community is remembered as a cause of disruptive quarrels ("drama"). To be fair to the furry community, there are those who say that this can be blamed on the zeal of furry haters as opposed to blaming it on the furry fandom: "Actually, some furry haters also start drama by whining as soon as they read the word 'furry'. I think this is also a reason to keep this stuff out of here." [11] Intolerance towards furry fandom may be self-amplifying.

The other point appears to be that opening the gates to furry material would lead to "furspam" overrunning the boards: "Furry subgenres would start their own threads and bump off other, more site-relevant content." [11] Given the large size of the furry community as opposed to the mostly obscure fetishes served on GUROchan, this could be a valid concern: Being confronted with fetishes not my own is okay as long as no single of these fetishes becomes overwhelming to what I am looking for. So let's segregate material with too overwhelming a community pressure behind it into its own spaces (furries have their very own imageboard site!) instead of letting it dominate all the minorities in a minority safe haven like the "Freakshow" board /f/. This may also be the reasoning behind segregating the "scat" fetish, immensely popular in its own right, into its own board. Unfortunately, this kind of segregation also keeps the audiences of the segregated fetishes isolated from the diversity found on a place like /f/. Minorities will only be visible to other minorities.

The most interesting thing about the ban on furry material, though, may be the difficulties it creates in separating non-furry material from furry material. A line of separation has to be drawn somewhere, and given the elaborate multitude of biologies on offer in /f/, a precise definition of furries as opposed to other human-animal hybrids, werewolves, girls with cat ears etc. is asked for. This is not an easy task, as /dis/ threads like "what is your definition of furry" [12] or "Yes to Beastiality, No to Furry?" [13] prove. But GUROchan's enthusiasm for such discussions brings

us closer to another matter: this site's intelligence in questions of body analysis and construction.

Deconstruction and reconstruction of bodies

I have talked a lot about the topical boards /s/, /g/ and /f/, in an amount that may appear unfair towards the other boards concerned exclusively with self-made art, requests and erotic story writing. But these three boards do provide orientation towards the main topical lines that also permeate the other boards: scat, gore, death and "weird"/"freakshow" subjects. I have explained that the scat board /s/ can be considered to be a sort of island isolated from the common diversity of /g/ and /f/, and I am going to mostly ignore it from this point on. Instead, I want to swim a bit more broadly through the common universe of /g/ and /f/, understanding that it expands to other boards as well.

Let's start in the gore board. /g/ is dedicated to destruction. Life processes are ended, bodily coherences are disintegrated. It's not just pure destructive madness towards bodies that can be found here, but

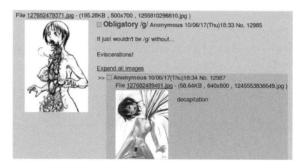

Typical GUROchan /g/ thread.
Thread URL: http://gurochan.net/g/res/12985.html

also anatomical curiosity. No body part, no matter how non-sexual, unspectacular or tiny, gets spared in the process of continuously deconstructing bodies in any way imaginable, separating skin from flesh and flesh from bone and bones from each other. Also, many ways to inflict pain, to attack the nervous system or to end life that are explored here confess an analytical interest in the workings of the human body and its experiences. How many organs (and in what order?) can I take out of the body during the vivisection of a human before they die, making the whole procedure much less enjoyable to both the surgeon and his victim? Is it possible to carefully impale a human non-mortally on a spit, and how long will they survive into the experience of getting spit-roasted before falling unconscious? Such are the questions negotiated in the "Gore / Death" category of GUROchan's porn, often quite seriously and scientifically, for certain seriousness creates a plausibility supporting the erotic experience. It helps that GUROchan seems to attract experts in medicine and anatomy, as a survey thread on /dis/ regarding actual or desired jobs of GURO-channers reveals. [14]

In contrast, /f/ is dedicated to the creation of bodies. Even in cases where the subject is a body reduced by amputation or magical disappearance of body parts, the interest lies not in the process of reduction but in exploring the situation of a new and unusual body image: What is it like to live without limbs, what erotic experiences can be created for such a being? Amputees here often provoke reconstructive efforts, being fitted with cyborg extremities, growing their bodily abilities even beyond their original, pre-amputation state, proving their intermediate reduction to be just a step in a larger transhuman transformation. In other cases, new bodies are created via inflation, magical growth of genitalia and being absorbed into other bodies. The fetish of being swallowed and digested alive called "vore" belongs on /f/ despite its expected ultimately deadly outcome for the victim: It's always depicted as being non-destructive, bloodless and painless, instead offering a unique and highly enjoyable erotic body experience for the person swallowed due to becoming intimately close to and joining the exotic body processes of their predator, often a reptile or fantasy creature. GUROchan /f/ celebrates the non-normal body, non- and trans-human eroticism, a death of bodily normativities.

That's the assembly line connecting /g/ and /f/: taking things apart analytically and putting them back together creatively. It's a sort of hacker approach

towards the sexual experience and the living body. I believe it has the potential to tear down bodily normativities like heterosexuality, health and species. Let me explain:

Throwing penises against heteronormativity

A frequent male experience towards "futanari" material is to find oneself masturbating to it and then wondering about one's heterosexuality: I thought I'm only sexually interested in the female form, so why do I jerk off to a being with an emphasized penis? [9]

Penises abound in mainstream pornography, including material aimed at heterosexual males. Somehow it's not supposed to be gay to find scenes involving penises erotic, as long as the penis targets a female form in an accepted heterosexual configuration – like penetrating a vagina. This configuration easily gets confused on a place like GUROchan's /f/ board, where body parts in general and genitalia in particular, unpinned thanks to the liberation gorily prepared on /g/, get stuck indiscriminately on anything that's around, including (but not limited to) people of the wrong gender.

Suddenly, the whole heteronormative logic breaks down. It was manly for a 100% anatomically male human to penetrate a 100% anatomically female human with a penis. But imagine a sex act of a penis protruding from the nipple of a lizard being's breast to anally rape a human hermaphrodite simultaneously masturbating his penis. How do you apply your heteronormativity to that? What's the more manly side? Or can this scenario be called homosexuality?

As more emphasis is put on fetishes alien to conventional concepts of sexuality, the polarities expected for these concepts matter less for the eroticism produced. If a certain type of gory bodily mutilation is requested on GUROchan, it's not unusual for gender to be totally irrelevant for the appreciation of the results. Whereas specific fetishes like furries or scat get segregated away on GUROchan, the idea

Example of (self-made) "shitting dicknipple" art, taken from GUROchan's /kaki/ board. Author "dicknips" writes:
"Yo Gurochan, i've got some new shitting dicknipples for ya. Coloring is kinda crappy but usually don't color my stuff. Pic is handdrawn, coloured in PS and here it is!"
Thread URL: http://gurochan.net/archives/kaki/res/8653. html

of segregating content by gender meets strong resistance; a predilection for a certain gender is treated as being on the same level as the other numerous fetishes, and "[p]opular fetishes don't each demand their own individual board", as one poster argues against gender-segregated boards. [15]

On GUROchan, biological sex is considered as one specific default configuration of organs and characteristics that ideally can be untied from each other and recombined with other organs and characteristics to create any new and unique configuration desirable, often outside of the spectrum of heteronormativity. If you're fascinated with penises for any reason and with large breasts as well, and with scat, put them together

on equal footing and create the genre of "shitting dicknipples".

A lot of threads on /dis/ express a cross-gender desire towards specific biological experiences only available to the other sex: Males ask what it is like to give birth and dream about being able to have an abortion [17], wonder how great it would be to masturbate with a vagina [16] and what having a period feels like [18]. At the same time, a man giving birth is a subject to be found graphically depicted on /f/. A much more popular genre depicts forced gender re-assignment, into both directions; unsurprisingly, the various shapings of hermaphroditism and transsexuality and possible medical methods of achieving them are also a popular topic on /dis/. [19]

Amputating health

Whereas GUROchan's "Gore / Death" board /g/ aims to be the very definition of unhealthiness, the "Freakshow" board /f/ can claim to totally ignore the polarity of "healthy" versus "unhealthy". The potency of a well-nourished, lively vore monster will get celebrated here just as much as anorexia.

Health defines norms and calls abnormalities unhealthy. A normal human being has two arms and two legs; someone with less would therefore be incomplete, unbalanced, problematical. Meet a thread where quadruple amputation is celebrated as a lifestyle, an ideal to attain, perfection instead of imperfection. Amputation – and quadruple amputation in particular – is one of the most popular fetishes on all of GUROchan (except /s/). GUROchan is one of many notorious corners on the web where people can affirm each other in their admiration of ideals considered unhealthy by most of the medical world, be it anorexia or "Body Integrity Identity Disorder" (longing for amputation of one's limbs).

A number of discussion threads carry titles like "Self Amputation" [20], "Amputation" [21], "How to really amputate a limb, as well as how to chemically blind someone…"[22], "Amputation of the Left Leg"[23] or "A little off the top"[24]. They do not deal primarily with fantasizing about how great it would feel to amputate another one's limbs or being amputated oneself (that happens in other threads and boards). They pose the actual question of how to practically get rid of one or more limbs in the real world, often stating outright the original poster's desire to have that happen in a real situation.

They do get answers. Those are hardly encouraging, often chiding the original posters as idiots if they really want to go through with it. But having once loudly disclaimed the intent of motivating someone's self-amputation, they usually continue with advice that does sound honest, sane and medically detailed (often involving not so much outright amateur amputation as producing a condition where professionals in a hospital will have no choice but to amputate); or by referring the original poster to GUROchan-external

Example of (self-made) amputee art, taken from GUROchan's /kaki/ board. Author "Magnet" writes:
"Hello~ My first guro pic, a somewhat plain sketch."
Thread URL: http://gurochan.net/archives/kaki/res/10092.html

Posted by Anonymous on GUROchan's /f/ board. One poster writes: "The artist is Umbrafox, that's part six of a series with Amy Pond."
Thread URL: http://gurochan.net/f/res/8404.html

sites of the body modification scene specialized in these questions.

Sure, it may be a most unhealthy intention to have one's arms or legs taken off without a medical need. But why should we expect people to care for healthiness that are obsessed with bodily destruction, pain and death? Injuries, deformities and sicknesses here are considered just one more module to configure new experiences of sex and body from. People asking "[What] awesome medical conditions"[25] "would you either a) like to have access to someone with, or b) like to have yourself?" can only marginally be expected to care about the normativities of health.

Dissolving species

As already mentioned the "Freakshow" board /f/ in particular is ripe with half- or non-human beings acting as sexual participants: animals, vore monsters, fairies, giants. Remembering a place like 4chan's /d/, this does not seem so unusual in itself: Apparently, anthropomorphization can make any object or creature acceptable as participant in scenarios reserved for humans in the mainstream of real life.

But the human form can hardly be called the standard or ideal that everything gravitates towards on /f/. Quite to the contrary: /f/ is dedicated to the dissolution of any ideal or standard of human form. The attraction of introducing non-human elements here is not so much in familiarizing them into something more human; it lies in further corrupting our idea of human exceptionalism by forcing it onto bodies it cannot safely digest without breaking or getting poisoned. Human-animal hybrids become uncanny if they look realistic instead of cartoony (cartoony like Furries are in their anthropomorphized cuteness), breaking up the barrier between us, God's image and the universe's only true intelligence and independent protagonist, and them, the dirty realm of ugly and dumb organisms, operated not by divine spirit and consciousness but by a nature entirely ignorant of our great humanistic virtues; a genealogy nominating worms and cockroaches as our mother, and dead matter as our grandmother.

Instead of pushing the spectator along a linear humanistic teleology, GUROchan in general and /f/ in particular hurl him chaotically around into many alternative directions that the human form could develop into, depriving a standardized "homo sapiens sapiens" of his privilege as a final ideal. This rule of morphological diversity and openness is accomplished by a fantasy world full of transformatory violence, medicine and magic. It is in stark contrast to any morphological standardization and idealization as it is achieved, for example, in Furry art. This may further help in explaining GUROchan's unwillingness towards Furries: Their style, threatening to establish a new standard and teleology similar to that of the human form ideal, may be incompatible with providing a sustainably open playground for diversified morphological tinkering.

This conflict is closely related with attacking the normativity of health. In the last decades, the discourse on medicine occupied itself with the

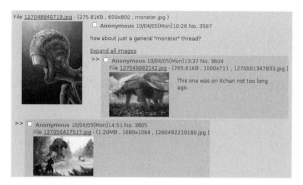

Typical, yet entirely non-humanistic and hardly pornographic GUROchan /f/ thread.
Thread URL: http://gurochan.net/f/res/3597.html

question of what a medicine with growing biotechnological abilities can be allowed to do: Is it to restrict itself towards repairing abnormalities back into a state deemed healthy by being normal, such as treating wounds and acquired diseases? Or is it to explore possibilities of improving organs and bodily functions beyond what would be "normal" for a human being (as dictated by natural evolution): providing limb prostheses stronger than ordinary limbs, doping endurance in sports or work, altering the genome to change skin color or lengthen lifetime? Conservatives want medicine to limit itself to ensuring a universal normativity of health and human body; progressives argue for an openness in changing the human body according to individual desire, transcending the original standard of the human form. This latter position falls under the label of "Transhumanism".

In attacking normativities of sex, health and even species, GUROchan's /f/ board can be considered an accomplice in this "transhumanist" cause. The idea of sex change may have been considered just as "grotesque" a few hundred years ago as most of the bodily transformations celebrated here; nowadays it is a medical procedure generally available and probably even covered by health insurance. A long path of changing sexual mores and questioning normativities, but also of rethinking the human body and its flexibilities, of tinkering with deconstructing and reconstructing it, was necessary to get from point A to point B. In my opinion, GUROchan provides a lot (though hardly enough to succeed by itself) of moral, aesthetic and intellectual work along those lines for alternative paths, from A to C or D or E, for reaching a growing variety of bodily forms, flexibilities and experiences in the future; one of the many beautiful projects porn can participate in.

References

1 http://fimoculous.com/archive/post-5738.cfm
2 http://www.4chan.org/rules
3 "The Dream of the Fisherman's Wife" is a famous Japanese drawing by Katsushika Hokusai, featuring an octopus having intercourse with a human female. It dates back to 1814. More recently, "tentacle porn" has become a sort of cliché for Japanese pornographic perversity in the western world.
4 http://www.4chan.org/news/?all#11
http://www.4chan.org/news/?all#47
5 http://en.wikipedia.org/wiki/Ero_guro
6 http://gurochan.net/news.php?p=rules
7 http://orz.gurochan.net/dis/archive/1188613974
8 http://orz.gurochan.net/dis/archive/1259373177

9 An example is this discussion thread about "Futa and faggotry": http://orz.guro-chan.net/dis/archive/1233756480
10 http://orz.gurochan.net/dis/archive/1242625066
11 http://orz.gurochan.net/dis/archive/1257888725
12 http://orz.gurochan.net/dis/archive/1239779423
13 http://orz.gurochan.net/dis/archive/1265939161
14 http://orz.gurochan.net/dis/archive/1266467682
15 http://orz.gurochan.net/dis/archive/1238639895
16 http://orz.gurochan.net/dis/archive/1179951604
17 http://orz.gurochan.net/dis/archive/1251290906
18 http://orz.gurochan.net/dis/archive/1252035149
19 See, for example, the thread "Dickgirls, Tgirls, Trannies, Shemales, Crossdressers and Traps": http://orz.gurochan.net/dis/archive/1179951604
20 http://orz.gurochan.net/dis/archive/1208501456
21 http://orz.gurochan.net/dis/archive/1235688323
22 http://orz.gurochan.net/dis/archive/1238807278
23 http://orz.gurochan.net/dis/archive/1240519166
24 http://orz.gurochan.net/dis/archive/1260319049
25 http://orz.gurochan.net/dis/archive/1232059904

Jonathon Keats

CINEMA BOTANICA
PORNOGRAPHY FOR PLANTS

Every day, all day long, we're awash in entertainment. Yet, while humans have reached saturation, other species have scarcely been approached as potential audiences. In particular, plants have been neglected despite attractive demographics, including a worldwide population many orders of magnitude greater than *Homo sapiens*.

I created *Cinema Botanica* to cater to trees and bushes and brambles, by providing video to suit their tastes and desires. Premiered in San Francisco at the Roxie Theater on the opening night of Arse Elektronika, *Cinema Botanica* took as its premise the natural allure of sex, the one subject certain to interest any species. Specifically, *Cinema Botanica* delivered pornography to the plant kingdom, screening explicit scenes of floral pollination by honeybees.

Because photosynthetic plants are sensitive to light but do not have the complex eyes of humans, *Cinema Botanica* employed specialized filmic techniques. The movie was shot entirely in light-and-shadow, and was projected directly onto the foliage with no sound. While botanical movie theaters in the future will undoubtedly go multiplex, the screening at the Roxie was quite modest: A digital projector streaming an endless loop of botanical smut onto the branches of half a dozen potted houseplants. People attending the Arse Elektronika premiere were able to see the video flickering on the plants' leaves, but of course their experience of the film was second-hand: *Cinema Botanica* was presented for the titillation of plants.

For Immediate Release

PORNOGRAPHY FOR HOUSE PLANTS PREMIERES IN SAN FRANCISCO

One-Night Screening at Roxie Theater on October 1st,... An Arse Elektronika Exclusive Engagement... Featuring Cinematography by Jonathon Keats...

September 14, 2009 - In a bid to increase movie audiences exponentially, and to dominate the motion picture industry, conceptual artist Jonathon Keats has announced plans to produce film and video for other species – from rose bushes to almond trees – using specialized new techniques. "Humans have more entertainment than they can endure," explains Mr. Keats. "Yet organisms with populations far greater than ours are routinely ignored by MGM and Disney."

Mr. Keats came to appreciate the potential impact of arts and entertainment on non-human audiences while choreographing ballet for honeybees at Yerba Buena Center for the Arts last year. "Dance comes naturally to bees," he says, "less naturally to trees. But all plants can perform photosynthesis. They're sensitive to the play of light. As an entertainment form, cinema was practically made for them."

By projecting specially-prepared video directly onto foliage, Mr. Keats found an effective way to share films with bushes and brambles, even entire forests and jungles. Still an essential question remained: What genres of film would appeal to flora? "This wasn't the sort of situation where I could learn the audience's mindset," admits Mr. Keats. "The only thing that would be a sure hit, I figured, was sex." Accordingly, the artist dutifully filmed plants getting pollinated, editing his uncensored footage into a gritty black-and-white porn video.

The film will premiere in San Francisco at the Roxie Theater on the opening night of Arse Elektronika, the world's leading sex and technology festival. In addition to the hundreds of human participants, a diverse population of house plants is expected to be in attendance.

"I think Cinema Botanica must be very titillating, if pollination is your thing," says festival organizer Johannes Grenzfurthner. Mr. Keats, who's already looking into further venues for plant porn, believes that the theater might even be intriguing to people. "Watching movies in a cineplex is partly about absorbing the experience of others in the audience. On the big screen, our point of view is enlarged. I see no reason why shared experiences with other species can't further expand our perspective."

BIO NOTES

monochrom

monochrom is a worldwide operating collective dealing with technology, art, context hacking, and philosophy which was founded in 1993. They specialize in an unpeculiar mixture of proto-aesthetic fringe work, pop attitude, subcultural science, and political activism. Their mission is conducted everywhere, but first and foremost 'in culture-archaeological digs into the seats (and pockets) of ideology and entertainment.'

Among their projects, monochrom has started Arse Elektronika, has released a leftist retro-gaming project, established a one baud semaphore line through the streets of San Francisco, started an illegal space race through Los Angeles, buried people alive in Vancouver, and cracked the hierarchies of the art system with the Thomann Project. They ate blood sausages made from their own blood in order to criticize the grotesque neoliberal formation of the world economy. Sometimes they compose melancholic pop songs about dying media and host the first annual festival concerned with cocktail robotics. They also do international soul trade, political sock puppet shows, aesthetic pregnancy counselling, food catering, and - sorry to mention - modern dance.

monochrom members

Johannes Grenzfurthner, Franz Ablinger, Harald List, Evelyn Fuerlinger, Frank Apunkt Schneider, Daniel Fabry, Guenther Friesinger, Anika Kronberger, Roland Gratzer.

www.monochrom.at/english/

Editors

Johannes Grenzfurthner is an artist, writer, curator, and director. He is the founder of monochrom. He teaches art theory and aesthetical practice at the University of Applied Sciences in Graz, Austria. Tag cloud: contemporary art, activism, performance, humor, philosophy, postmodernism, media theory, cultural studies, science fiction, and the debate about copyright.

Guenther Friesinger lives in Vienna and Graz as a philosopher, artist, writer, curator and edu-hacker. He is founder and head of the paraflows festival, co-founder and head of the QDK - quarter for digital culture, member of monochrom, co-organizer of the Arse Elektronika festival and the Roboexotica festival. Publications: Mind and Matter: Comparative Approaches towards Complexity, Urban Hacking: Cultural Jamming Strategies in the Risky Spaces of Modernity, Public Fictions, Pr0nnovation?: Pornography and Technological Innovation. Tag cloud: contemorary art, activism, media theory, radical innovation, science fiction, free software and copyright.

Daniel Fabry is designer, researcher and lecturer in the lofty fields of media and interaction design, working at the University of Applied Sciences in Graz, Austria. He is multiartist and polyatheist and member of monochrom.

Saul Albert is an artist and technologist from London whose work emerged from the intersection of 'net art, DIY culture and the Free Software movement in the 90's and continues to develop forms of participatory culture, technology and governance.

In 2006 he co-founded (with Michael Weinkove) The People Speak (http://theps.net), to create 'tools for the world to take over itself' and get people to talk to each other.

Thomas Ballhausen studied Comparative Literature and German at the University of Vienna. He is a lecturer at the University of Vienna and at the University of Applied Arts, head of the Studies-Department of the Austrian Film Archive, and has authored publications on film history, media theory, and popular culture.

Tatiana Bazzichelli is a communication sociologist and expert in network culture, hacktivism and net art. She is PhD Scholar at the University of Aarhus in Denmark (Information and Media Studies) and Visiting Researcher at the Stanford Humanities Lab, Stanford University. She wrote the book Networking. The Net as Artwork published by the Digital Aesthetics Research Center of Aarhus, 2009 (www.networkingart.eu/english.html). Her fields of interest are networking and hacktivism as a strategy for art, and in 2001 she founded the networking project AHA:Activism-Hacking-Artivism (www.ecn.org/aha), which won the Honorary Mention for Digital Communities at the Ars Electronica Festival (Linz, 2007). She organized, together with Gaia Novati, CUM2CUT: Indie-Porn-Short-Film-Festival (2006-2008), an independent porn competition first based in Berlin and further developed as a nomadic entity (www.cum2cut.net). Towards the end of the 90's she organized several exhibitions and conventions on media art and hacktivism, such as HACK.Fem.EAST (www.hackfemeast.org, Berlin, 2008), HackMIT! (Berlin, 2007), Hack.it.art (Berlin 2005), Art on the Net in Italy (Berlin 2005), MediaDemocracy and Telestreet (Munich, 2004), AHA (Rome, 2002).

Violet Blue is pro blogger, podcaster, vlogger and femmebot at Metblogs SF, Geek Entertainment TV, and Gawker Media's Fleshbot. And sex columnist for the San Francisco Chronicle, and a Forbes Web Celeb. She is a best-selling, award-winning author and editor of almost two dozen books; some translated into five languages.

Micha Cárdenas / dj lotu5 / Azdel Slade is a transgender artist, theorist and trouble maker. She will be a Lecturer in the Visual Arts department at UCSD in Fall and Winter of 2009. She is an Artist/Researcher in the Experimental Game Lab at CRCA and the b.a.n.g. lab at Calit2. Her interests include the interplay of technology, gender, sex and biopolitics. She blogs at Transreal.org. Micha holds an MFA from the University of California San Diego, an MA in Media and Communications with distinction from the European Graduate School and a BS in Computer Science from Florida International University. Micha is a a curator and collective member at the Lui Velazquez space in Tijuana. She has exhibited and performed in Los Angeles, San Diego, Tijuana, New York, San Francisco, Montreal, Egypt, Ecuador, Spain, Colombia and many other places. Micha has received grants from UCIRA, calit2 and Ars Virtua and her work has been written about in publications including the LA Times, San Diego Union Tribune, .dpi magazine and Rolling Stone Italy.

Kim De Vries earner her MA and PhD at the University of Massachusetts Amherst in English (Rhetoric and Composition) and has been teaching rhetoric and writing since 1994. Her research interests include global rhetorics, new/digital media, and internet culture. Recently she has been studying the institutionalization of new media, particularly the involvement of women, and is conducting a series of interviews to gather material on how gender shapes our engagement with technology. A member of the polyamorous community, she is exploring how gender, technology, and social networks intersect in the formation and empowerment of a marginalized group. You can find her work at her blog, "Else-If-Then" which can be found at www.kdevries.net/blog/. Comments are very welcome!

Robert Glashuettner is a radio journalist, presenter and writer based in Vienna/Austria. His specialises in video game culture, cyberculture, 8-bit-music and fashion. He finished communication studies at the University of Vienna, holds a BA in Audio Engineering and finished second place at the Pac-Man World Championships in New York City.

Christian Heller lives in Berlin and experiments with the outsourcing of identity and thought processes in internet software under the pseudonym "plomlompom."

Katrien Jacobs is a scholar, curator and artist in the field of new media and sexuality and works at The Chinese University of Hong Kong. She was born in Belgium and received her Ph.D. degree in comparative literature and media from the University of Maryland, with a thesis on dismemberment myths and rituals in 1960s/1970s body art and performance media. She has organized netporn conferences in recent years with the Institute of Network Cultures. She published Libi_doc: Journeys in the Performance of Sex Art (2005, Maska Publications), Netporn: DIY Web Culture and Sexual Politics (Lanham: Rowman and Littlefield, 2007) and is currently working a book about Chinese pornography and Internet culture for Intellect Books. Her work can be found at www.libidot.org

Jonathon Keats, acclaimed as "a poet of ideas" by the New Yorker, is an experimental philosopher, artist, and writer based in the United States and Italy. Recently he choreographed a ballet for honeybees at Yerba Buena Center for the Arts. He has also exhibited extraterrestrial abstract artwork at the Judah L. Magnes Museum, unveiled a prototype ouija voting booth for the 2008 election at the Berkeley Art Museum, and attempted to genetically engineer God in collaboration with scientists at the University of California. Exhibited internationally, his projects have been documented by PBS, NPR, and the BBC World Service, garnering favorable attention in periodicals ranging from the San Francisco Chronicle and the Los Angeles Times, to Nature and New Scientist, to Flash Art and ArtUS. Since graduating summa cum laude from Amherst College in 1994, he has been a visiting artist at Chico State University, Montana State University, and UC Davis, and has been awarded fellowships by Yaddo, the MacDowell Colony, the Ucross Foundation, the MacNamara Foundation, and the Poetry Center at the University of Arizona. He is represented by Modernism Gallery in San Francisco.

Monika Kribusz is artist, writer, performer and tantric. She is working in the fields of gender, mystic, consciousness expanding techniques, she is a researcher in the ancient roots of Indian tantrism. Monika is working on the integration of body-mind-spirit developing techniques in postmodern reality. In her art projects she is hacking the male heterosexuality dominated common perception of spirituality. She encourages the individual to create their own identity in adventurous and exciting ways. She uses the laboratory of applied sexuality to discover new ways of interaction between human beings. She uses modern technologies to show the flow of energy in the body during sexual intercourse as proof of what the tantric tradition discovered thousands of years ago. Polyamory, bisexuality, sexuality used as a generator for human creative energy are her actual topics.

Kyle Machulis, aka qDot, is a researcher of alternate input mechanisms and haptics, which is really a fancy way of saying he breaks sex toys. Through his Slashdong webpage (http://www.slashdong.org), he uses the topic of teledildonics (remotely actuated sexual experience) to teach the basic concepts of software, electrical and mechanical engineering. He also tracks the convergence of sex and technological advances in toys and interaction, building on the idea that paradigms for interfaces people would use for intimate encounters on computers can be extended to other usage experiences.

Elle Mehrmand is a performance/new media artist and musician who uses the body, electronics, video, photography, sound and installation within her work. She is the singer and trombone player of Assembly of Mazes, a music collective who create dark, electronic, middle eastern, rhythmic jazz rock. Elle is currently an MFA candidate at UCSD, and received her BFA in art photography with a minor in music at CSULB. She has received grants from UCIRA and Fine Arts Affiliates. She is a researcher at CRCA and the b.a.n.g. lab at UCSD. Elle is a member of the Lui Velazquez gallery collective in Tijuana. Her performances have been shown in Long Beach, Los Angeles, Colombia, Tijuana, Montreal, Dublin and San Diego.

Pepper Mint is a San Francisco organizer, activist, and social theorist in the bisexual, polyamory, and BDSM movements. He holds a Master's in Computer Science from MIT, where he started doing queer activism in 1994. With his partner Jen, he puts on nonmonogamy workshops in the San Francisco area, and he is currently running three different local polyamory events. In addition, he moderates various internet polyamory forums and has been active in online poly communities since 2001. Pepper has written a number of papers and essays that approach polyamory, BDSM, and bisexuality from a queer theory perspective, including a paper on cheating published in the Journal of Bisexuality. To view his writings, visit www.pepperminty.com and www.freak-sexual.com. He would love to hear from you at pepomint@gmail.com.

Annalee Newitz is a writer who covers the collisions between technology and media, culture and science. She writes for many periodicals from Popular Science to Wired, and since 1999 has had a syndicated weekly column called Techsploitation. From 2004-2005 she was a policy analyst for the Electronic Frontier Foundation. She is the editor of io9, a blog about science fiction owned by the Gawker Media Network.

Ani Niow, pink, sexy and occasionally averse to pants, San Francisco based, has decided in lieu of a proper bio, you should grab the nearest dictionary and look up "awesome". In summary... Ani Niow spends her time hacking sex toys, volunteering for sex-positive feminist minded art galleries such as Femina Potens, and helping to put on the monthly 5ive Minutes of Fame at SF's Noisebridge hackerspace. Rising to infamy in 2009 by creating the steampunk vibrator, Ani hopes to create an entire line of mechanically interesting sex toys in the near future.

Rainer Prohaska lives and works in Vienna, Austria. Experiments with modular temporary sculptures as architectonic interventions in public space and with mobile objects. In this experiments the "Performative Act of Constructing" and the "Effects of the Public Space" on this process play a crucial role. The construction methods of these sculptures and objects are inspired by toy kits like Lego and Matador. [E.g.: Toy-Kit Architectures , "The 'Z'-Boats", "Enter Beijing"] // Cross-Media works that deal with entanglements of real space and virtual space. The orchestration of these projects works with methods of contemporary performing art. These methods are described in Rainer Prohaska's "C.O.H.R.-Theory" as the "Construction of Hybrid Realities". [E.g.: Operation Cntrcpy, KRFTWRK] // Artistic concepts, which take on common processes, transform and present them as a modified reality in performances and fine art projects. One focus is "Cooking as a basis for Fine Art and Performing Art Works". [E.g.: Restaurant Transformable].

Carol Queen, Ph.D. curates the antique vibrator museum at Good Vibrations, where she also serves as Staff Sexologist and Chief Cultural Officer. She is the founding director of the Center for Sex & Culture, a sex education, library, archive, and cultural center in San Francisco. She has written three books (Exhibitionism for the Shy, Real Live Nude Girl, and Firecracker Alternative Book Award winner The Leather Daddy and the Femme) and edited (or co-edited) eight collections of erotica and personal essays, including the Lambda Literary Award-winning PoMoSexuals. She's a frequent commentator on television and in documentary films about sex, and has appeared in several sex ed videos, including explicit cult classic Bend Over Boyfriend: a Couple's Guide to Male Anal Pleasure. She uses her academic perspective as a sociologist and cultural sexologist to view contemporary and historical society and understand the roles sexuality plays and has played. Please visit her: www.carolqueen.com, www.goodvibes.com, www.sexandculture.org, www.carolqueenblog.com.

Bonni Rambatan is an independent cultural researcher, theorist, and blogger. His primary field of research is the role of technology in shaping contemporary human subjectivity, sexuality, and society. Rambatan aims to develop a critical approach to society and politics with his unique blend of Lacanian psychoanalysis and informatics and new media studies, interrogating our experience with the computer monitor and how it alters our notion of subjectivity and our relations to local and global politics and the market. Rambatan's blog can be found at http://posthumanmarxist.wordpress.com.

Eleanor Saitta is a designer, artist, hacker, and researcher working at the intersections between mediums ranging from interaction design and architecture through jewelry and fashion, with an emphasis on the seamless integration of technology into lived experience and the humanity of objects and the built environment. She has previously worked at the NASA Ames Research Center and the IBM Almaden Research Center, and divides her time between Seattle and New York, as life allows.

Randy Sarafan is a new-media artist who is currently the Director of Imagineering for SF Media Labs, a Virtual Fellow with F.A.T. (Free Art and Technology) Lab, and a top project contributor at Instructables. com, the leading user-generated how-to website. Randy's work has a unique personal style and inherent ingenuity that can only be described as "troubling" by any reasonably sound individual. He has graduated with honors in the Design Technology program at Parsons School of Design and lives in San Francisco.

Jack Sargeant is a writer specialising in cult film, underground film, and independent film, as well as subcultures, true crime, and other aspects of the unusual. In addition he is a film programmer and an academic. He currently lives in Australia.
Recent publications: 'In Celebration of Going Too Far: Waterpower' in From the Arthouse to the Grindhouse, eds Rob Weiner & John Cline, Scarecrow, 2010; 'This Is Hardcore' in Cinematic Folds: The Furling and Unfurling of Images, ed Firoza Elavia, Pleasure Dome, 2008; 'Revealing and Concealing: Notes and Observations on Eroticism and Female Pubic Hair', in Hair Styling, Cultur and Fashion, eds Geraldine Biddle-Perry & Sarah Cheang, Berg, 2008; 'Harry Crosby', Straight to Hell: 20th Century Suicides, ed. Namida King, London: Creation Books, 2004.

R. U. Sirius (born Ken Goffman) is a US writer, editor, talk show host, musician and cyberculture icon, best known as co-founder and original Editor-In-Chief of Mondo 2000 magazine from 1989–1993.

Douglas Bryan LeConte-Spink is a co-founder of Baneki Privacy Computing (baneki.com) - a no-compromise provider of world-class network security and privacy services – and is a longtime activist in the field of inter-species sociocultural symbiosis. Currently, he is, in his own words: "a political prisoner within the U.S. prison system; as a result of my longtime academic interest in reciprocal models of human/non-human emotional bonding, social connectivity, and inter-species cooperation - as well as my longstanding work in the areas of free-speech, anti-censorship, and customer-friendly privacy/encryption technologies - I have been targeted by ideologues within the federal criminal justice system. I was sentenced in 2010 to 3 years' imprisonment... despite being neither charged with - nor convicted of – any alleged crime, in order to 'teach me a lesson about respecting authority.' Appeals are fully in-process." Further information on Mr. LeConte-Spink's campaign against bigotry is available at cultureghost.org. Mr. LeConte-Spink carries an MBA from the University of Chicago, a B.A. in cultural anthropology from Reed College, and has studied complex systems theory at the doctoral level. He is a fixed-object jumper who has opened many new exit points worldwide (BASE 715), a practicing Zen Buddhist, a successful mentor to several International-level show jumping stallions, a longtime technology entrepreneur, a former operational member of a US/Canadian helicopter smuggling crew, an organizer of underground electronic music gatherings in the Pacific Northwest, and served proudly as a front-line activist for Earth First! during the Old Growth wars of the early 1990s. Currently, he pursues his academic interests as an independent researcher, having published extensively in numerous fields. He is the founder of the Deep Symbiosis Institute (information available soon at deepsymbiosis. org), Exitpoint Stallions Limitee (www.exitpoint. org), and is a founding member of the Deep Justice Network (deepjustice.net). He can be contacted via wrinko@hushmail.com.

Allen Stein is a co-founder of Seattle based 'RandD' firm SenseTrak which is focused on technology innovations applied to sensuality training. His insatiable curiosity in emerging technologies and his expertise in business development lead him to invent the world's first Internet-controlled commercial sex machine, thethrillhammer, and the underlying technology infrastructure for mass consumer deployment.

The initial popularity of thethrillhammer propelled his work into designing various luxury pleasure crafts and sex devices for a very discrete private clientele and adult industry clients. The machines feature the latest in sexual technology and everything from back massagers, auto lube, or the often requested vaporizers, all under Internet control. To fund his research in human computer interaction sexual with machines, Allen began producing AVN Award-nominated scenes for the adult industry leaders, including Homegrown Video and Pink Visual. Twice nominated he was crushed when he lost Best Solo Scene to I Love Big Toys Volume 2. He is forever worried that he will be a Susan Lucci of porn.

Uncle Abdul a Professional Electrical Engineer with nearly 50 years of experience and comes with a proven expertise in the elecrical sexual stimulation field. In 1998 under the nom de plume, Uncle Abdul, "JuiceElectricity for Pleasure and Pain" was written. It has subsequently has become the premier primer on the subject of electrical stimulation for sexual purposes (otherwise known as E-Stim). Unc' has also presented on this very topic around the US and in Canada on numerous occassionsincluding being an invited guest lecturer for the Wardell B. Pomeroy Lecture Series to health care professionals at San Francisco's Institute for Advanced Study of Human Sexuality. Unc' also have a website on the subject (see http://www.UncleAbdul.com) providing supplemental information on E-Stim. He is the world acknowledged expert in this field.

Noah Weinstein is a traveler, pig roaster, pizza enthusiast, dreamer and general doer of things. He is currently the Research Director at SF Media Labs, a whitewater raft guide with ARTA (American River Touring Association), and Head of Production at Instructables.com. His previous work includes, social research, custom speaker building, exhibit creation and taxidermy. Noah graduated with honors from the Environmental Studies and Visual Art departments at

Brown University and currently lives in Oakland, CA. He has helped teach 1.2 million people how to kiss.

Rose White is a sociology PhD student at Graduate CenterCUNY, in Manhattan, but lives in Brooklyn because that's where all the action is. Her doctoral work is on how bright people break rules (in other words, deviance and technology!)- tell her what you want studied and maybe she'll get on that. In 1997 she had a story anthologized in Best American Erotica, which means it's about time for her to write some new porn.

RE/Search Catalog

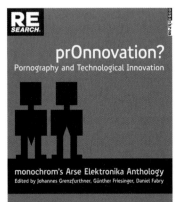

PUNK & D.I.Y.

PUNK '77: an inside look at the San Francisco rock n' roll scene, 1977 by James Stark

Covers the beginnings of the S.F. Punk Rock scene through the Sex Pistols' concert at Winterland in Jan., 1978, in interviews and photographs by James Stark. James was among the many artists involved in early punk. His photos were published in *New York Rocker, Search & Destroy* and *Slash*, among others. His posters for Crime are classics and highly prized collectors' items. Over 100 photos, including many behind-the-scenes looks at the bands who made things happen: Nuns, Avengers, Crime, Screamers, Negative Trend, Dils, Germs, UXA, etc. Interviews with the bands and people early on the scene give intimate, often darkly humorous glimpses of events in a *Please Kill Me* (Legs McNeil) style. "The photos themselves, a generous 115 of them, are richly satisfying. They're the kind of photos one wants to see..."—Puncture. "I would recommend this book not only for old-timers looking for nostalgia, but especially to young Punks who have no idea how this all got off the ground, who take today's Punk for granted, to see how precarious it was at birth, what a fluke it was, and to perhaps be able to get a fresh perspective on today's scene needs..."—MAXIMUMROCKNROLL 7½x10¼", 98 pp, 100+ photos, on archival art paper. PB, **$20**

SEARCH & DESTROY: The Complete Reprint (2 big 10x15" volumes)
"The best punk publication ever"—Jello Biafra
Facsimile editions (at 90% size) include all the interviews, articles, ads, illustrations and photos. Captures the enduring revolutionary spirit of punk rock, 1977-1978. Vol. I contains an abrasive intro-interview with Jello Biafra on the history and future of punk rock. Published by V. Vale before his RE/Search series, *Search & Destroy* is a definitive, first-hand documentation of the punk rock cultural revolution, printed as it happened! Patti Smith, Iggy Pop, Ramones, Sex Pistols, Clash, DEVO, Avengers, Mutants, Dead Kennedys, William S. Burroughs, J.G. Ballard, John Waters, Russ Meyer, and David Lynch (to name a few) offer permanent inspiration and guidance. First appearance of Bruce Conner Punk photos. 10x15", 148pp, Only 200 copies left of Volume Two: **$19.95 (Volume One is long out-of-print)**

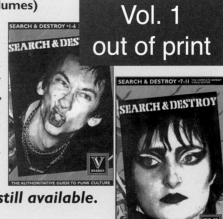

Some of the Search & Destroy original tabloid issues are still available. Please call or email for information.

INDUSTRIAL CULTURE HANDBOOK DELUXE HARDBACK

This book is a secret weapon—it provided an educational upbringing for many of the most radical artists practicing today! The rich ideas of the *Industrial Culture* movement's performance artists and musicians are nakedly exposed: *Survival Research Laboratories, Throbbing Gristle, Cabaret Voltaire, SPK, Non, Monte Cazazza, Johanna Went, Sordide Sentimental, R&N, & Z'ev*. **Topics include:** brain research, forbidden medical texts & films, creative crime & *interesting* criminals, modern warfare & weaponry, neglected gore films & their directors, psychotic lyrics in past pop songs, and *art brut*. Limited Edition of only 1000 hardbacks on glossy paper.
8½x11", 140 pp, 179 photos & illust. PB, **$40**

ZINES Vol. 1 & 2 The Punk Rock Principle of "DO-IT-YOURSELF" (D-I-Y) inspired the creation of "ZINES": handmade self-publications by creative individuals on topics ideally against status quo thinking and ideas. Zines such as Murder Can Be Fun, Beer Frame, Crap Hound, Thrift Score, Bunny Hop, Fat Girl, Housewife Turned Assassin gleefully show the satisfactions to be had by "publishing it yourself." These two books will inspire and provoke readers to become publishers, and have been used in college classes as textbooks. Both books heavily illustrated with photographs and illustrations from Zinemakers' lives & works. **Zines 1:** 184 pp, PB, **$18.99. Zines 2:** 148 pp, PB, **$14.99. *GET BOTH for $20.00 (plus shipping for two items).***

BODY MODIFICATION AND S&M

RE/Search 12: MODERN PRIMITIVES The *New York Times* called this "the Bible of the underground tattooing and body piercing movement." *Modern Primitives* launched an entire '90s subculture. Crammed with illustrations & information, it's now considered a classic. The best texts on ancient human decoration practices such as tattooing, piercing, scarification and more. 279 eye-opening photos and graphics; 22 in-depth interviews with some of the most colorful people on the planet. "Dispassionate ethnography that lets people put their behavior in its own context."—*Voice Literary Supplement* "The photographs and illustrations are both explicit and astounding ... provides fascinating food for thought. —*Iron Horse* 8½x11", 212 pp, 279 photos and illus, PB. Great gift! **$19.50**

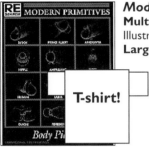

Modern Primitives T-shirt!
Multi-color on black 100% cotton T-shirt
Illustrations of 12 erotic piercings and implants. **Xtra**
Large only. Dare to Wear It! $16.

SPECIAL OFFER: MODERN PRIMITIVES
BOOK & T-SHIRT GIFT-PACK—ONLY $29.00
PLUS SHIPPING.

Confessions of Wanda von Sacher-Masoch Married for 10 years to Leopold von Sacher-Masoch (author: *Venus in Furs* & many other novels) whose whip-and-fur bedroom games spawned the term "masochism," Wanda's story is a feminist classic from 100 years ago. She was forced to play "sadistic" roles in Leopold's fantasies to ensure the survival of herself & their 3 children–games which called into question who was the Master and who the Slave. Besides being a compelling story of a woman's search for her own identity, strength and, ultimately, complete independence, this is a true-life adventure story–an odyssey through many lands peopled by amazing characters. Here is a woman's consistent unblinking investigation of the limits of morality and the deepest meanings of love. "Extravagantly designed in an illustrated, oversized edition that is a pleasure to hold. It is also exquisitely written, engaging and literary and turns our preconceptions upside down."—*L.A. Reader* 8½x11", 136 pp, illustrated, PB. **$20**

The Torture Garden by **Octave Mirbeau** This book was once described as the "most sickening work of art of the nineteenth century!" Long out of print, Octave Mirbeau's macabre classic (1899) features a corrupt Frenchman and an insatiably cruel Englishwoman who meet and then frequent a fantastic 19th century Chinese garden where torture is practiced as an art form. The fascinating, horrific narrative slithers deep into the human spirit, uncovering murderous proclivities and demented desires. "Hot with the fever of ecstatic, prohibited joys, as cruel as a thumbscrew and as luxuriant as an Oriental tapestry. Exotic, perverse ... hailed by the critics."—Charles Hanson *Towne* 8½x11", 120 pp, 21 mesmerizing photos. **PB: $25. Rare Hardcover (edition of only 100; treat yourself!): $40**

Bob Flanagan Super Masochist Born 1952 and deceased in 1996, Bob grew up with Cystic Fibrosis, and discovered extended S&M practices as a secret, hand-picked pathway towards life extension. In flabbergastingly detailed interviews, Bob described his sexual practices and his relationship with long-term partner and Mistress, the artist Sheree Rose. Through his insider's perspective we learn about branding, piercing, whipping, bondage and ingenious, improvised endurance trials. Includes photographs by Sheree Rose. This book "inspired" a movie, which used many of the questions found in this book. "...an elegant tour through the psychic terrain of SM." -- Details Magazine 8½x11", 136 pp, illustrated, PB. **$20.00**

PRANKS

RE/Search 11: PRANKS! A prank is a "trick, a mischievous act, a ludicrous act." Although not regarded as poetic or artistic acts, pranks constitute an art form and genre in themselves. Here pranksters challenge the sovereign authority of words, images and behavioral convention. This iconoclastic compendium will dazzle and delight all lovers of humor, satire and iron. "I love this book. I thought I was the only weirdo out there, but this book inspires me to be weirder. I pick it up weekly, even though I've read it from covr to cover many times. Still cracks me up." (reader) "The definitive treatment of the subject, offering extensive interviews with 36 contemporary tricksters including Henry Rollins, Abbie Hoffman, Jello Biafra, SRL, Karen Finley, John Waters, DEVO . . . from the Underground's answer to Studs Terkel."—*Washington Post* 8½x11″, 240 pp, 164 photos & illustrations, PB, **$25, Deluxe HB $40.**

PRANKS 2 "*Pranks woke me up from a deep slumber. It's as if a demolition crew had a party in my brain.*" *(reader)* An extended underground of surrealist artists like The Suicide Club, Billboard Liberation Front, DEVO, John Waters, Lydia Lunch & Monte Cazazza give inspiring tales of mirth and conceptual mayhem. Includes Internet pranks, art pranks, prank groups and more! 8'x10″, 200 pp,, many photos & illustrations, PB, glossy paper, 8x10″, **$20.**

Mr Death T-Shirt by ManWoman. Black & yellow on White 100% heavyweight cotton T-Shirt. Sizes: S,M,L,XL. Limited edition, 50 copies: **$25.**

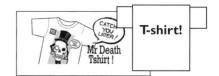

TWO BY DANIEL P. MANNIX

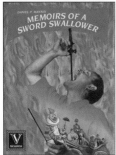

MEMOIRS OF A SWORD SWALLOWER Not for the faint-of-heart, this book will GROSS SOME PEOPLE OUT and delight others. "I probably never would have become America's leading fire-eater if Flamo the Great hadn't happened to explode that night . . ." So begins this true story of life with a traveling carnival, peopled by amazing characters—the Human Ostrich, the Human Salamander, Jolly Daisy, etc.—who commit outrageous feats of wizardry. One of the only *authentic* narratives revealing the "tricks" (or rather, painful skills) involved in a sideshow, and is invaluable to those aspiring to this profession. OVER 50 RARE PHOTOS taken by Mannix in the 1930s.
8½x11″, 128 pp, 50+ photos, index, PB, **$20**
Signed copies available for only $30

FREAKS: We Who Are Not As Others

Amazing Photos! This book engages the reader in a struggle of wits: Who is the freak? What is normal? What are the limits of the human body? A fascinating, classic book, based on Mannix's personal acquaintance with sideshow stars such as the Alligator Man and the Monkey Woman. Read all about the notorious love affairs of midgets; the amazing story of the Elephant Boy; the unusual amours of Jolly Daisy, the fat woman; hermaphrodite love; the bulb-eating Human Ostrich, etc. **Put this on your coffee table and watch the fun!** 8½x11″, 124 pp, 88 photos. PB. **$20** Author died in 1997. **Signed, hardbound copies available for $50**

J.G. BALLARD

J.G. Ballard: Quotes Amazing, provocative quotes from J..G. Ballard illuminating the human condition, arranged by topic. Edited by V. Vale with Mike Ryan. Dozens of gorgeous photos by Ana Barrado, Charles Gatewood and others. "Ballard understands the transformation technology can effect on human desire." (Observer)
ISBN 1-889307-12-2, 416 pages, index, 5″ x 7″, gloss paper, **$20.** Limited AUTOGRAPHED Flexibind Edition of only 250 copies, signed by J.G. Ballard himself, only **$75.** Also, unsigned Library Flexibind Editions (only 100 printed) available, only **$35.**

J.G. Ballard Conversations
British luminary, J.G. Ballard converses with V. Vale, Mark Pauline, Graeme Revell, David Pringle and other forward thinkers. Photographs by Ana Barrado, Charles Gatewood and others. Some topics: Sex, technology, the future, plastic surgery, child-raising, *Empire of the Sun*. Index, book recommendations, and interviews are also included.
ISBN 1-889307-13-0, 360 pages, index, 5″ x 7″, gloss paper, **$20**

RE/Search 8/9: J.G. Ballard J.G. Ballard is our finest living visionary writer. His classic, *CRASH* (made into a movie by David Cronenberg) was the first book to investigate the psychopathological implications of the car crash, uncovering our darkest sexual crevices. He accurately predicted our media-saturated, information-overloaded environment where our most intimate fantasies and dreams involve pop stars and other public figures. Intvs, texts, critical articles, bibliography, biography.
 Also contains a wide selection of quotations. "Highly recommended as both an introduction and a tribute to this remarkable writer."—*Washington Post* "The most detailed, probing and comprehensive study of Ballard on the market."—*Boston Phoenix*.
8½x11″, 176 pp, illus. PB. **$20 (last copies; order soon!)**

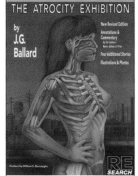

Atrocity Exhibition A dangerous imaginary work; as William Burroughs put it, "This book stirs sexual depths untouched by the hardest-core illustrated porn." Amazingly perverse medical illustrations by Phoebe Gloeckner, and haunting "Ruins of the Space Age" photos by Ana Barrado. Our most beautiful book, now used in many "Futurology" college classes. 8½x11″, 136 pp, illus. PB **$20.**
LIMITED EDITION OF SIGNED HARDBACKS $150 (Only 20 copies left)

W.S.BURROUGHS

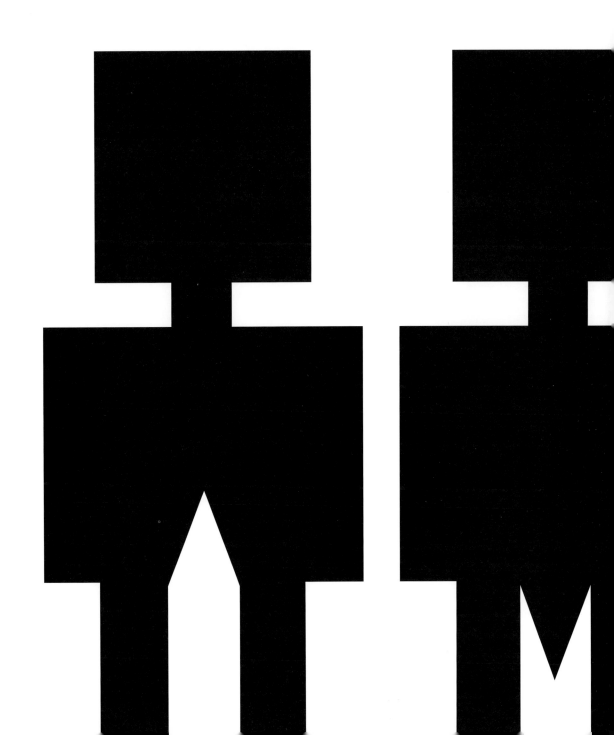

monochrom Catalog

Do Androids Sleep with Electric Sheep?

Critical Perspectives on Sexuality and Pornography in Science and Social Fiction

J. Grenzfurthner, G. Friesinger, D. Fabry, T. Ballhausen (eds.)

ISBN: 978-1889307237   RE/Search / edition mono/monochrom   (264p.)

This anthology stands under the motto "future" – and the ways in which the present sees itself reflected in it. Maintaining a broadened perspective on technical development and technology while also putting special emphasis on its social implementation, this anthology focuses on Science and Social Fiction.

prOnnovation?

Pornography and Technological Innovation

Johannes Grenzfurthner, Guenther Friesinger, Daniel Fabry (eds.)

ISBN: 978 1889307206   RE/Search / edition mono/monochrom   (200p.)

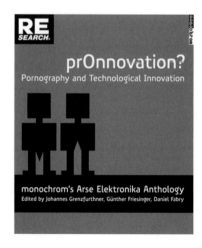

From the depiction of a vulva in a cave painting to the newest internet porno, technology and sexuality have always been closely linked. No one can predict what the future will bring, but history indicates that sex will continue to play an essential role in technological development. Is it going too far to assume that research in nanotechnology and genetic engineering will be influenced by our sexual needs? The question is not whether these technologies alter humanity, but how they do so.

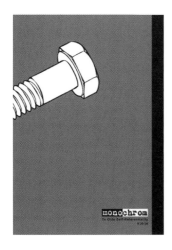

monochrom #26-34

Ye Olde Self-Referentiality

monochrom

ISBN 978-3-9502372-6-9   edition mono/monochrom (500p.)

monochrom is a magazine object appearing in telephone book format, which is published by the art/tech group of the same name. monochrom came into being in the mid-1990s as a fanzine for cyberculture, science, theory, cultural studies and the archaeology of pop culture in everyday life. Its collage format is reminiscent both of the early DIY fanzines of the punk and new wave underground and of the artist books of figures such as Dieter Roth, Martin Kippenberger and others. With a great deal of forced discontinuity, a cohesive potpourri of digital and analog subversion is pressed between the covers of monochrom. Each issue is an unnostalgic amalgam of 125 years of Western counterculture cocked, aimed and ready to fire at the present. It is a Sears catalog of subjective and objective irreconcilability – the Godzilla version of the conventional coffee table book.

Mind and Matter

Comparative Approaches towards Complexity

Guenther Friesinger, Johannes Grenzfurthner, Thomas Ballhausen (eds.)

ISBN 978-3-8376-1800-6   [transcript] (230p.)

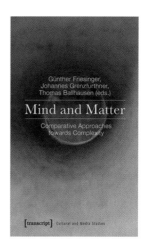

The terms »mind« and »matter« appear to signify two concepts irreplaceable and permanent in nature. The increasing challenges and modes of reflection of digital life and cultural creation have contributed to a productive doubting of said dichotomy. Net culture has exposed the causality of the two only superficially contradictory systems and translated these into new technological realities. This publication, using an interdisciplinary approach, strives to investigate the entanglement of cultural, artistic and technical praxis, to document the developments, to clarify the status quo of the scientific community in a practical and exemplary fashion and to enable glimpses of potential future developments.

Urban Hacking

Cultural Jamming Strategies in the Risky Spaces of Modernity

Guenther Friesinger, Johannes Grenzfurthner, Thomas Ballhausen (eds.)

ISBN 978-3-8376-1536-4   [transcript] (230p.)

Urban spaces became battlefields, signifiers have been invaded, new structures have been established: Netculture replaced counterculture in most parts and also focused on the everchanging environments of the modern city. Important questions have been brought up to date and reasked, taking current positions and discourses into account. The major question still remains, namely how to create culturally based resistance under the influence of capitalistic pressure and conservative politics. This collection of essays and contributions attempts to address this question and its implications for different scientific and artistic fields.

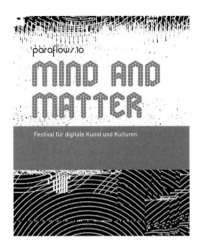

paraflows 10 - Mind and Matter
Annual Convention for Digital Arts and Cultures / Vienna – Catalogue
Guenther Friesinger (ed.)
ISBN: 978-3-9502372-9-0   edition mono/monochrom (96p.)

paraflows 10 – Mind and Matter seeks to put on display how hard- and software, content and object in contemporary art not only co-exist, but also condition and shape each other. What role does hardware play for media infrastructure – how is the net set up, what are the real pillars on which virtuality rests? How does the sculptural process look like, from a technological point of view? What are the effects of technological prospects on physical representation? What are the challenges that digital art must face, when it comes to conserving its works? Mind and body are inextricably tied to each other, yet we are tempted to regard mental processes as independent of their bodily basis.

paraflows 09 - Urban Hacking
Annual Convention for Digital Arts and Cultures / Vienna – Catalogue
Guenther Friesinger (ed.)
ISBN: 978-3-9502372-5-2   edition mono/monochrom (110p.)

The city of Vienna has a long tradition of interventions and actions in the public space. With this year's topic Urban Hacking, the fourth paraflows festival continues this artistic investigation of the public and urban space of living. What is more, paraflows 09 will shed light on the role played by digital media when it comes to exploring, questioning, and shaping the urban infrastructure.

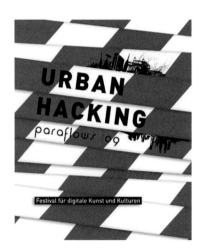

paraflows 08 - Utopia
Annual Convention for Digital Arts and Cultures / Vienna – Catalogue
Guenther Friesinger, Judith Fegerl (eds.)
ISBN: 978-3-9502372-2-1   edition mono/monochrom (112p.)

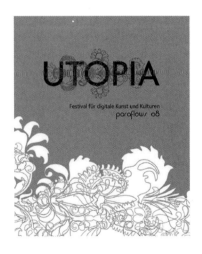

Usually, technology plays a major role in utopian phantasms. More often than not their realistic implementation is a matter of availability of technology and very often it is – especially in digital art – the starting point of a work of art. paraflows 08 – Utopia tries to develop the perspective, the linguistic roots of which corresponds with last year's exhibitions. In the context of the topic Utopia, it gathers concepts of a possible future, draws perspectives, dreams and prognoses, dares taking a prognostic look at the future.

Ten Years of Roboexotica

G. Friesinger, M. Wurzer, J. Grenzfurthner, F. Ablinger, C. Veigl (eds.)

ISBN: 978-3-9502372-3-8   edition mono/monochrom (120p.)

ROBOEXOTICA takes a look at the broadly interpreted field of (modern) cocktail culture along with a variety of arts contributing to it. Until the year 1999 there were no attempts to create a competition of developing technologies or to introduce cocktail robotics as an index for the integration of technological achievements in everyday life and as a means of documenting the creation of new interfaces for man-machine-interaction, a priori dedicated to hedonism. For a whole decade now that void is filled by the ROBOEXOTICA, the Viennese cocktail robotics festival around which a pool of creative people has gathered who, year after year, are actually providing new ideas on the topic.

monochrom: Carefully Selected Moments

Audio CD, 17 tracks, 74 min.

Hello there! I am the old salt lamp that nobody likes. So I was very happy when the Vienna-based art group monochrom asked me if I would like to be on the cover of their Greatest Hits album 'Carefully Selected Moments'. monochrom aren't a regular band. They don't have time to get involved in stupid band stuff like owning instruments or rehearsing. That's why they asked some friends if they could help record these songs. Friends such as German house legend Hans Nieswandt, Matthias Kertal (from Mika), Gerald Votava, GameJew, Der Schwimmer, Max of Prey/audiospam, Gegenstimmen and the non-existing average Viennese laptop musician Tonki Gebauer. They all said "Well, yes, that sounds like the kind of collaboration that makes this world a place you can recommend to your children". And so they did. And they did so in a way that still makes me wonder how they did it. It's a marvellous collection of songs and moments. Well, legal moments, that is. Stupid copyright asshole law!

1349 Mission St.

CENTER FOR
SEX & CULTURE

Sex Education
Arts & Culture
Events, Meetings
Library and Archive

☎ 415-902-2071
Info@SexandCulture.org
Mailing Address
2261 Market St. #455A
SF, CA 94114

1349 Mission St.

CSC
CENTER FOR
SEX & CULTURE

Sex Education
Arts & Culture
Events, Meetings
Library and Archive

☎ 415-902-2071
Info@SexandCulture.org
Mailing Address
2261 Market St. #455A
SF, CA 94114

1349 Mission St.

CENTER FOR
SEX & CULTURE

Sex Education
Arts & Culture
Events, Meetings
Library and Archive

☎ 415-902-2071
Info@SexandCulture.org
Mailing Address
2261 Market St. #455A
SF, CA 94114

1349 Mission St.

CENTER FOR
SEX & CULTURE

Sex Education
Arts & Culture
Events, Meetings
Library and Archive

☎ 415-902-2071
Info@SexandCulture.org
Mailing Address
2261 Market St. #455A
SF, CA 94114

1349 Mission St.

CENTER FOR
SEX & CULTURE

Sex Education
Arts & Culture
Events, Meetings
Library and Archive

☎ 415-902-2071
Info@SexandCulture.org
Mailing Address
2261 Market St. #455A
SF, CA 94114

1349 Mission St.

CSC
CENTER FOR
SEX & CULTURE

Sex Education
Arts & Culture
Events, Meetings
Library and Archive

☎ 415-902-2071
Info@SexandCulture.org
Mailing Address
2261 Market St. #455A
SF, CA 94114

1349 Mission St.

CENTER FOR
SEX & CULTURE

Sex Education
Arts & Culture
Events, Meetings
Library and Archive

☎ 415-902-2071
Info@SexandCulture.org
Mailing Address
2261 Market St. #455A
SF, CA 94114